Lecture Notes in Electrical Engineering 1042

The book series *Lecture Notes in Electrical Engineering* (LNEE) publishes the latest developments in Electrical Engineering—quickly, informally and in high quality. While original research reported in proceedings and monographs has traditionally formed the core of LNEE, we also encourage authors to submit books devoted to supporting student education and professional training in the various fields and applications areas of electrical engineering. The series cover classical and emerging topics concerning:

- Communication Engineering, Information Theory and Networks
- Electronics Engineering and Microelectronics
- Signal, Image and Speech Processing
- Wireless and Mobile Communication
- Circuits and Systems
- Energy Systems, Power Electronics and Electrical Machines
- Electro-optical Engineering
- Instrumentation Engineering
- Avionics Engineering
- Control Systems
- Internet-of-Things and Cybersecurity
- Biomedical Devices, MEMS and NEMS

For general information about this book series, comments or suggestions, please contact leontina.dicecco@springer.com.

To submit a proposal or request further information, please contact the Publishing Editor in your country:

China

Jasmine Dou, Editor (jasmine.dou@springer.com)

India, Japan, Rest of Asia

Swati Meherishi, Editorial Director (Swati.Meherishi@springer.com)

Southeast Asia, Australia, New Zealand

Ramesh Nath Premnath, Editor (ramesh.premnath@springernature.com)

USA, Canada

Michael Luby, Senior Editor (michael.luby@springer.com)

All other Countries

Leontina Di Cecco, Senior Editor (leontina.dicecco@springer.com)

** **This series is indexed by EI Compendex and Scopus databases.** **

Xiaoqing Zeng · Xiongyao Xie · Jian Sun ·
Limin Ma · Yinong Chen
Editors

Proceedings of the 5th International Symposium for Intelligent Transportation and Smart City (ITASC)

Branch of the International Symposium on Autonomous Decentralized Systems (ISADS) 2023

Springer

Editors
Xiaoqing Zeng
College of Transportation Engineering
Tongji University
Shanghai, China

Jian Sun
College of Future Transportation
Chang'an University
Xi'an, Shaanxi, China

Yinong Chen
School of Computing and Augmented Intelligence
Arizona State University
Tempe, AZ, USA

Xiongyao Xie
College of Civil Engineering
Tongji University
Shanghai, Shanghai, China

Limin Ma
College of Environmental Science and Engineering
Tongji University
Shanghai, Shanghai, China

ISSN 1876-1100 ISSN 1876-1119 (electronic)
Lecture Notes in Electrical Engineering
ISBN 978-981-99-2251-2 ISBN 978-981-99-2252-9 (eBook)
https://doi.org/10.1007/978-981-99-2252-9

This Springer imprint is published by the registered company Springer Nature Singapore Pte Ltd.
The registered company address is: 152 Beach Road, #21-01/04 Gateway East, Singapore 189721, Singapore

Preface

The symposium of International Symposium on Intelligent Transportation and Smart City (ITASC), was originated from several experts' idea of communicating the technology of transportation and IT. The first symposium (ITASC 2013) was held in Mexico in 2013, when main technologies discussed in that symposium have been built and operated in rail transit in China and Japan. Since then, the ITASC symposium was held every two years, and the latest ITASC in 2022 attracted 16 top scholars and nearly 400 people from both universities and corporations. The proceedings of ITASC 2019 and ITASC 2022 attracted specialists and scholars from both universities and corporations who came from Asia, America and Europe.

ITASC 2022 was held on May 20–21 in Shanghai in 2022. Despite the delay due to Covid-19 pandemic, ITASC committee still created a chance for many people to share latest research results. Topics on developments in transportation, civil engineering, IT and environment engineering were discussed in this symposium by online meeting. Now we can keep focusing on the theories and applications of modern technologies of intelligent transportation and smart city, theoretically combine them together, and so as to create better city life. Moreover, the symposium also provided a chance for corporations to get new achievements, and then new techniques and products could be soon put into transportation, municipal construction and information controlling, and thus, promote the progress of the technology in these domains.

In this proceedings, we selected 20 articles from more than 60 submitted articles, which cover several professional fields, such as transportation, civil and environmental engineering. Quite a few of these articles chose specific prospective to describe new ideas of crossing different professional fields, which will give readers great inspiration. As the prosperous urban development in China, there are urgent needs and huge application prospects to keep more and more attention on the intelligent transportation and smart city, which actually has already been successfully implemented in some developed areas.

We hope all those interested in improving our urban living qualities could attend our symposium to share your wonderful discoveries and experiences, no matter you are from universities, entrepreneurs or governments. We believe our symposium will benefit more and more participants and create wider effect. Here, we give our sincere thanks to all sponsors, press, print and electronic media for their excellent coverage of this symposium.

Xiaoqing Zeng

Contents

About the Editors

Prof. Xiaoqing Zeng is a professor at College of Transportation Engineering, Tongji University, Shanghai, China. She received her Bachelor, Master and Ph.D. degrees from Tongji University. Her research focuses in the field of transportation information and control engineering. She has published more than 80 papers, received 28 patents, wrote 6 books, one of which was awarded the first prize by Tongji University. Prof. Zeng has rich teaching experience. She teaches 18 undergraduate and graduate courses, such as Rail Transit Control and Management, Traffic Information Systems and Engineering, ATC Technology of Urban Mass Transit, etc. Prof. Zeng is also the Secretary General of Shanghai Creative Studies Institute.

Prof. Xiongyao Xie is a professor at College of Civil Engineering, Tongji University, China, and deputy director of Department of Geotechnical Engineering. He received Ph.D. degree in Underground Construction and Engineering at Tongji University, in 2001. His research focuses on NDT, risks and disaster prevention of tunneling and underground engineering. He has published more than 40 papers, received 4 patents, completed nearly 20 research projects and has been awarded several national prizes in China. Prof. Xie also has rich teaching experience, and his course "Underground Structures and Underground Engineering Construction" was awarded the excellent courses prize by Tongji University.

Prof. Jian Sun is a professor at the College of Future Transportation, Chang'an University, China. He received Ph.D. degree in Traffic and Transportation Engineering from University of Florida, 2009. Prof. Sun's research field includes urban driving behavior and environment, urban comprehensive traffic planning and simulation, vehicle networking technology and urban traffic signal self-adaption. He has published nearly 50 papers in the past five years, participated in 20 research projects and had 7 patents.

Prof. Limin Ma is a professor at the School of Environmental Science and Engineering, Tongji University, China. He obtained Ph.D. degree in Biogeochemistry from Institute of Earth Environment, Chinese Academy of Sciences, in 2001. Prof. Ma worked as a postdoc in the College of Environmental Science and Engineering, Tongji University, from 2001 to 2003. From 2009 to 2010, he was visiting scholar in Pisa University, Italy. In the past years, his research includes biogeochemistry, past environmental change, theory and technology of restoration of the degraded ecological ecosystem and degraded environment, control of pesticide pollution, theory and technology of constructed wetland for wastewater ecological treatment. He has published more than 100 scientific papers and been granted 7 patents.

Prof. Yinong Chen is a full teaching professor at School of Computing and Augmented Intelligence, Arizona State University, Tempe, USA. He received Ph.D. from the University of Karlsruhe, Germany, in 1993. He did postdoctoral research at Karlsruhe and at LAAS-CNRS in France. From 1994 to 2000, he was a lecturer and then senior lecturer in the School of Computer Science at the University of the Witwatersrand, Johannesburg. Dr. Chen joined Arizona State University in 2001. His expertise includes service-oriented computing, programming languages, dependable computing, robotics and embedded systems. He (co-) authored 10 textbooks and more than 300 research papers. He is in editorial boards of several journals, including Journal of Systems and Software and Simulation Modeling Practice and Theory.

Traffic Simulation and Autonomous Driving Experiment in VIPLE

Yinong Chen(✉) and Gennaro De Luca

School of Computing and Augmented Intelligence, Arizona State University, Tempe, AZ, USA
{yinong.chen,gennaro.deluca}@asu.edu

Abstract. This paper presents the development of visual programming languages and summarizes the recent research in traffic simulation and autonomous driving experiments in VIPLE (Visual IoT/Robotics Programming Language Environment). The traffic simulator is developed in the Unity programming engine. The traffic simulator allows generation of different traffic patterns, either randomly or guided through a recorded real-world traffic dataset. Traffic experiments can be programmed in VIPLE or in Unity simulator, allowing developers to create such experiments without a deep understanding and lengthy support programming around the environment. Instead, the user can focus on implementing optimal routing algorithms that navigate a vehicle through the city traffic. The autonomous driving environment is based on TORCS simulation environment. A variety of difficulty levels of autonomous driving experiments can be created. Both traffic simulation and autonomous driving experiments can be written in visual programming language VIPLE, as well as in C# and Python within VIPLE environment. VIPLE has been widely used in different courses worldwide. This new advancement in VIPLE can particularly help students to perform traffic control and autonomous driving related programming, machine learning, and artificial intelligence research and experiments.

Keywords: Traffic simulation · dynamic routing · autonomous driving · computer science education

1 Introduction

Software engineering has evolved in several generations. The first generation was from the 1970s to 1980s, where the focus was to use structured programming languages to construct well-structure computer programs. The second generation of software engineering was from the 1990s to 2000s, where design patterns, modeling, and object-oriented programming were introduced. The third generation of software engineering was from the 2000s to 2010s, where architecture-driven development, service-oriented computing, cloud computing, and visual programming languages were introduced. The fourth generation of software engineering is based on big data processing, machine learning, and hardware-software co-design [1].

As a part of the third generation of software engineering, visual programming languages emerged in the 2000s, which allow software compositions through connecting

X. Zeng et al. (Eds.): ITASC 2022, LNEE 1042, pp. 1–13, 2023.
https://doi.org/10.1007/978-981-99-2252-9_1

existing components and services using orchestration and choreography styles [1, 2]. In the orchestration composition style, a central process takes control over the involved components and services and coordinates the execution of different operations. This composition style is defined by the following features.

- The involved components and services communicate with the central process only, within the application.
- The overall functionality is achieved by aggregating other components and services.
- The composition style is particularly useful for private business process, using independent components and services.
- BPEL (Business Process Execution Language) is the flagship language using this style.
- A complementary composition is choreography, which is defined by the following features.
- There is no central coordinator.
- Each component and service involved can communicate with multiple partners within the application.
- Each component and service decide how to interact with other services to consume their functionality.
- The composition style is particularly useful for public business process involving coordinated design of distributed services.
- WS-CDL is an example language using this style.

The main applications of visual programming environments and the composition styles are in business and in education, as visual programming environments help business developers and students to develop applications without deep understanding of computer organizations and programming language details at low level, such as syntax design, memory management, resource sharing, process synchronization, etc. [3]. As software engineering enters its fourth generation, some visual programming languages also include big data processing and machine learning capabilities. VIPLE is one such visual programming languages that has included fourth generation software engineering features [4].

In this paper, we introduce different visual programming environments used in business and in education. We focus on discussing the new features that are recently developed in VIPLE programming environment, including traffic simulation with machine learning capacity and autonomous driving, and quantum computing.

The basic parts of VIPLE have been used at different grade levels in middle and high schools. The advanced parts, such as traffic simulation, autonomous driving, and quantum computing are being used at senior and graduate levels at universities [5].

The rest of the paper is organized as follows. Section 2 gives an overview of the composition languages and their composition styles. Section 3 discusses VIPLE visual components and composition examples. Section 4 presents the recent development and new functions in VIPLE and their applications in traffic simulation, autonomous driving. We also briefly introduce quantum computing for machine learning. Finally, Sect. 5 concludes the paper.

2 Development of Visual Programming Languages

This section gives an overview of different visual programming language environments.

Visual programming language environments wrap components and services into visual blocks and enable drag and drop methods to compose workflows and business processes. The basic concepts of visual programming languages were discussed, and the existing attempts are surveyed in early research by Berkely scientists [6]. Another survey of visual programming languages was conducted by Mishra in 2017 [7]. New languages with practical applications were developed since 2000. Some of these languages are commercial products and others are widely used in computer science education. Figure 1 shows the major composition languages and environments that support visual composition of workflows and business processes.

There are full released and well documented visual programming languages. There are many more that are for research only. PSML-S was a research language that helped to establish the initial research of visual programming language development at ASU [8].

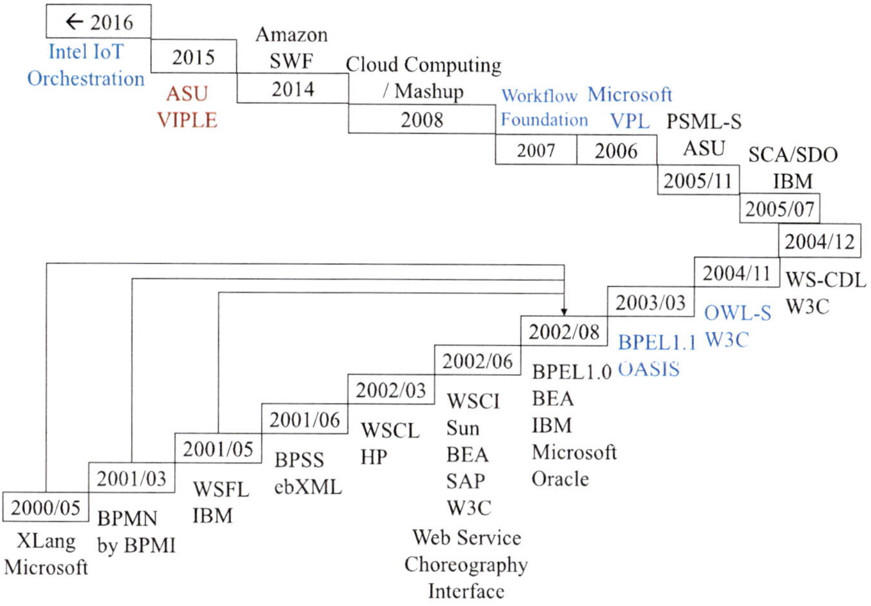

Fig. 1. Development of Visual composition languages and environments

Table 1 lists the architecture styles and the features of a few key languages that are used in business process composition. These languages are either visual notations themselves, such as BPMN [9] and WF [10], or have external visual tools to visualize the text notations, such as BPEL [11]. Some of these languages are standards defined by standard organizations, and these languages can have different implementations by different vendors, such as BPEL and BPMN. Some of the languages are proprietary to certain companies. For example, WF is part of Visual Studio by Microsoft, and SWF a part of Amazon Web Services cloud [12]. As indicated in Fig. 1, BPEL is defined by different companies jointly based on XLang, BPMN, and WSFL to make sure that it is not vendor dependent. It has become a major standard support across the industry.

Table 1. Key languages and features in business applications

Language environment	Architecture Style	Other features
BPEL 2002	Orchestration	Based on SOAP, WSDL, UDDI directory, widely used by large corps
BPSS 2001 BP Spec. Schema	Choreography	Business Process Specification Schema Based on SOAP, ebXML repository, CPP/CPA collaboration, for small biz
BPMN 2001 BP Modeling Notation	Orchestration	A superset of BPEL, supports advanced semantics & complex structures, by BPMI which merged with OMG in 2005
WSCI 2002	Choreography	WS Choreography Interface: Complementary to BPEL, submitted to W3C, not widely used
WS-CDL 2004	Choreography	Complementary to BPEL, W3C own proposal
Work-flow Foundation WF 2007	Orchestration and Choreography	A general programming environment allowing visual workflows and code activities for general computing and business processes
SWF 2014	Orchestration	Used for connecting Amazon lambda-functions and APIs to form business process in AWS cloud

In addition to business process oriented visual languages, there are a set of visual programming languages used for education. These languages allow students without programming background to implement their logic ideas and build their computational thinking concepts. These languages are used at different grade levels. Some languages are created by universities for public access, such as Scratch [13], Alice [14], App Inventor [15], and VIPLE [1, 3, 4]. Some are proprietary products associated with certain hardware products, such as Microsoft VPL [16] and Lego EV3 [17].

Table 2 lists a few key languages in this category, including their main purposes and grade level of students who can start to use the languages.

Table 2. Languages for computing education

Language	Main purpose	Starting from
MIT Scratch	Visual game and movie programming	Primary school and up
Alice	Visual game and movie programming	Middle school and up
MIT App Inventor	Phone App visual programming	High school and up
Lego EV3	Lego EV3 robot programming	Middle school and up
Microsoft VPL	Robot programming	High school and up
Intel IOT SOL	IoT and embedded system programming	College students
ASU VIPLE	General programming and robot programming	High school and up

3 VIPLE

VIPLE is a visual programming language environment designed for general-purpose programming. It particularly offers activities and services for IoT and robotics application building. This section reviews the general-purpose computing part, and the next section will present the recent new features of VIPLE.

As a general-purpose programming language, VIPLE offers the necessary programming constructs, such as assignment, calculate, if-then-else, and loop. Figure 2 shows the basic activities and general-purpose services that allow the construction of general-purpose programs.

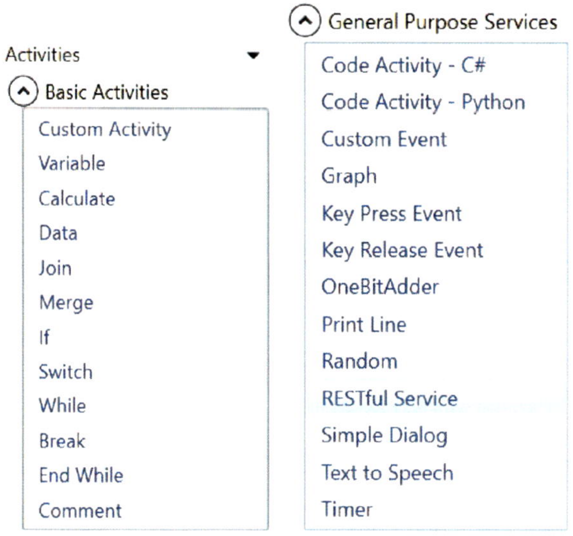

Fig. 2. VIPLE basic activities and general-purpose services

Figure 3 shows a VIPLE program that counts from 0 to 10. It uses Text to Speech service to read the current number and uses Print Line service to print "All Done" at the end of counting. A Merge basic activity is used for selecting the initial entrance of the loop and the reentrance of the loop. A variable Counter is used to store the increasing number, and a Calculate activity is used to perform add 1 operation.

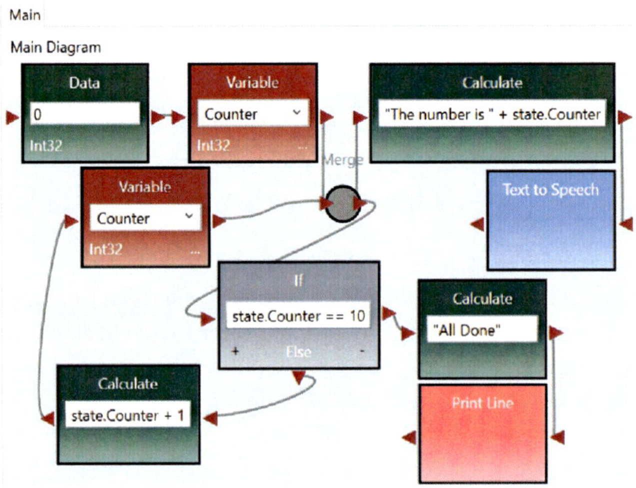

Fig. 3. A VIPLE program that counts from 0 to 10

The same program can also be implemented using a While loop activity.

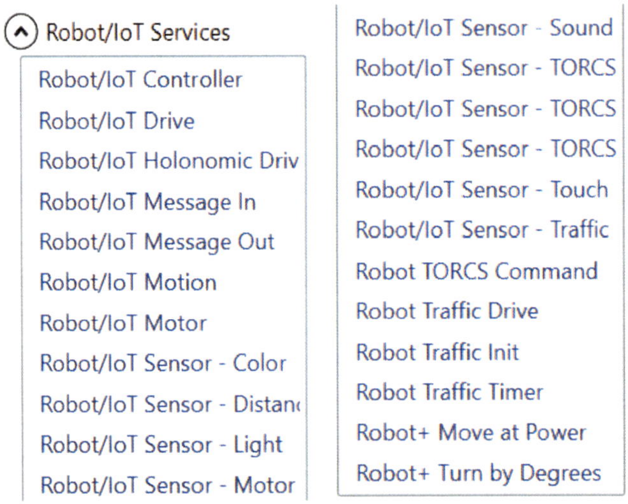

Fig. 4. VIPLE Robot/IoT services

VIPLE is called an IoT and robotics programming language, because it offers a large set of Robot/IoT services. Figure 4 shows the list of such services.

There are four sets of different types of services.

Robot/IoT Controller is the main service that defines a control center of a Robot/IoT system. One or more controllers can be used, specifying a single control center or multiple control centers.

The next set of services includes the Robot/IoT sensor, drive, and message services. These services will be associated with one of the controllers to give the controller input or to take output from the controller to perform actions.

The third set of services are TORCS services for autonomous driving experiments, which will use TOCS simulator to exhibit the driving simulation. We will discuss this set of services in the next section.

The last set of the services included is the traffic simulation services that will communicate with a Unity Traffic Simulator. We will discuss this set of services in the next section.

Figure 5 shows the commands that can start different functions.

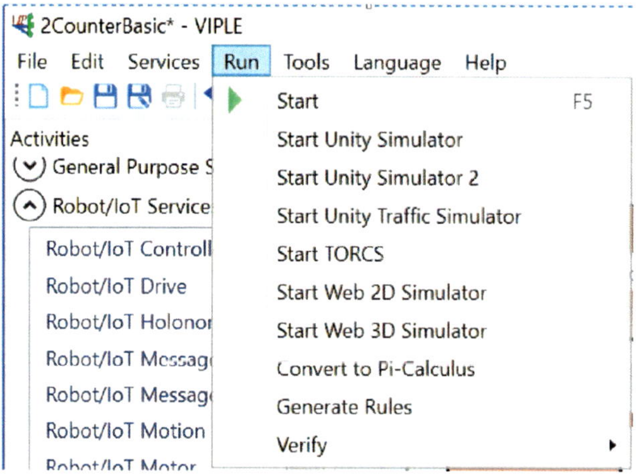

Fig. 5. VIPLE Start commands

The first command Start will start the execution of a VIPLE program. If a simulator is involved, we need to start the simulator first before we start the VIPLE program.

The next three Start commands will start one of the three Unity simulators embedded into VIPLE. We will discuss Start Unity Traffic Simulator in the next section.

The next Command is to start TORCS simulator [18].

The next two commands are to start the Web 2D maze simulator and to start the Web 3D maze simulator, respectively. These simulators are developed within VIPLE environment [1].

In addition to the simulators, VIPLE also supports Pi-Calculus based program analysis, proof, and verification [19, 20], which are shown in the next three commands in Fig. 5.

4 Latest Additions to VIPLE

This section presents the latest developments in VIPLE, including traffic simulation, autonomous driving, and quantum computing.

4.1 Traffic Simulation

A traffic simulator is recently developed using Unity game engine. The simulator will generate a city map and random or controlled traffic (cars marked in yellow) on the streets. Once started the simulation, a red car will appear in the map, and it can be programmed in VIPLE to travel from the current position to any position on the map. Figure 6 shows the overall map with clusters of dense areas connected by freeways.

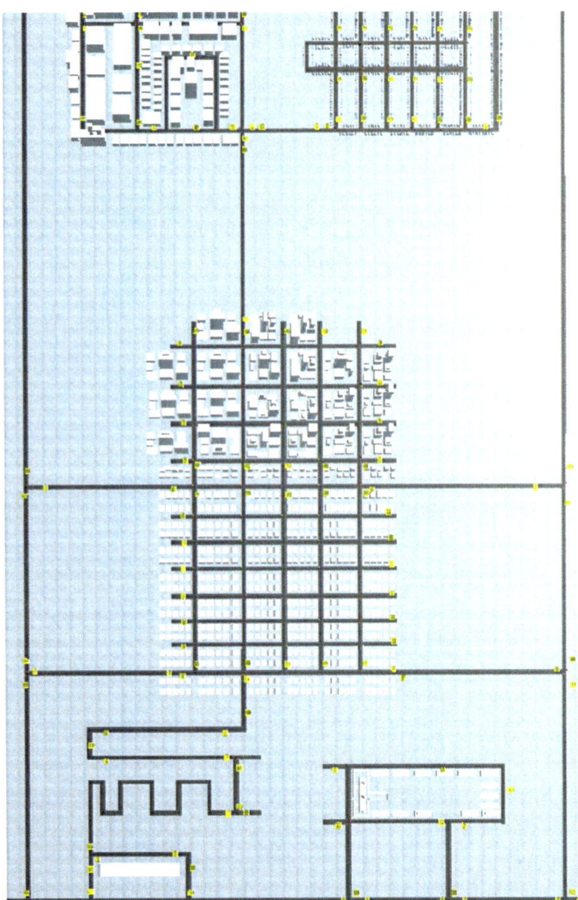

Fig. 6. The overall traffic map

Figure 7 shows a part of the map with the red car, multiple yellow cars, and the controls on the map. The controls on the map can define the traffic density (number of yellow cars in the map), destination of the red car, planning path and trace of the red car, showing different areas of the city, different views of the map, zoom in and zoom out, etc.

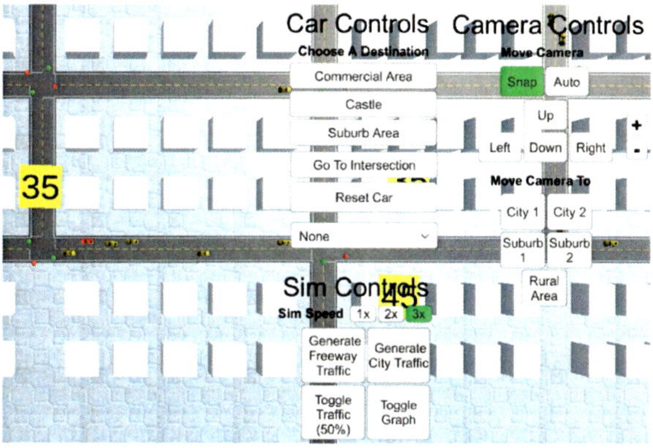

Fig. 7. Traffic map showing controls and cars in streets

The map contains traffic lights, traffic signs (e.g., stop sign), speed limits on each section of streets, distance of each section of streets, etc. These pieces of information are sent to VIPLE in real time and VIPLE can program the red car to implement the shortest-distance path to the destination and the shortest-time path to the destination.

The traffic on the streets can be generated randomly or guided by know traffic patterns. In the latest study, we have used Arizona Maricopa government published data to guide our traffic generation and write VIPLE program to find the shortest-time path [21]. A video demonstrating the traffic simulation with dynamic shortest distance computing in the simulation environment is available at [23].

4.2 Autonomous Driving

Another major development in VIPLE is the TORCS autonomous driving experiments. VIPLE makes autonomous driving experiments simple to write and easy to understand. Figure 8 shows a simple VIPLE program. The main code is in the Code Activity in C#, as shown in Fig. 9.

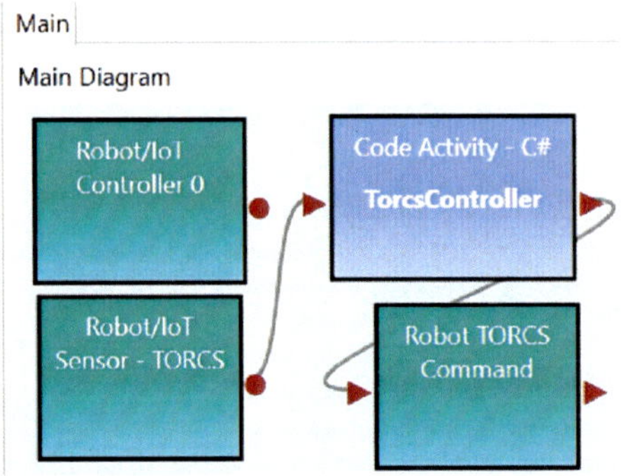

Fig. 8. Simple VIPLE code for autonomous driving

The code keeps the car in the middle of the racetrack, maintaining acceleration at 0.3, driving slowly at gear 2, and not applying brake at all. The car can complete the racetrack in about 74 s.

```
public override void Execute()
{
  // The values are passed in as an object, need to unbox it.
  var inputDictionary = Input as Dictionary<string,object>;

  // Returning a dictionary to reference the values by name.
  var toReturn = new Dictionary<string, double>()
  {
    {"accel", 0.3},
    {"brake", 0.0},
    {"gear", 2},
    // Need to cast these values to doubles
    {"steer", (double)inputDictionary["angle"]-
              (double)inputDictionary["trackPos"]*0.5},
    {"clutch", 0.2}
  };
  Output = toReturn;
}
```

Fig. 9. Code Activity in the VIPLE code

Obviously, this program is not optimal to completing the racetrack. In a real race program, we need to raise the gear as the speed increase, maximize the acceleration when the track is straight, and need to brake before the curve. An optimized VIPLE program is shown in Fig. 10.

This program can complete the racetrack in 54 s, reducing from 74 s.

VIPLE also supports two-car racing, where each car also needs to sense the other car to avoid the collision. Figure 11 shows the experiment with two cars in the same racetrack. We can ally the same algorithm or two different algorithms for the two cars.

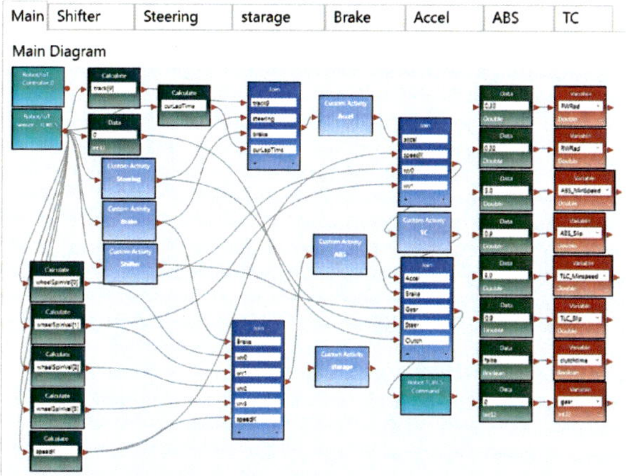

Fig. 10. Optimized VIPLE code for autonomous driving

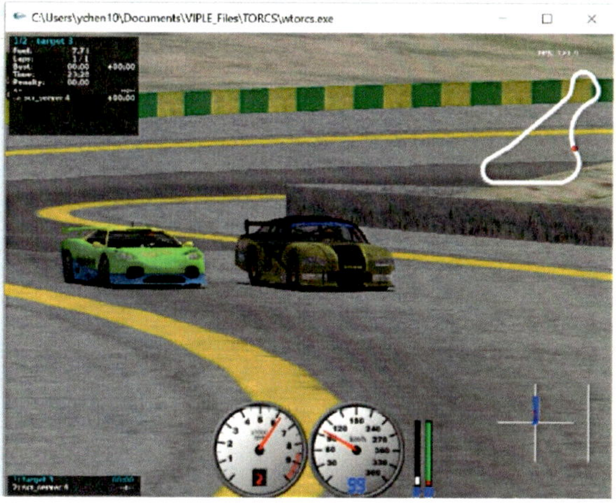

Fig. 11. Two car races in the same racetrack

4.3 Quantum Computing

The latest development in VIPLE is the inclusion of quantum computing, particularly for accelerating machine learning [22]. Figure 12 shows Quantum Basic Activities and Quantum Services that have been built in VIPLE.

From these activities and services, different computationally intensive applications can be developed, particularly for accelerating machine learning training and testing.

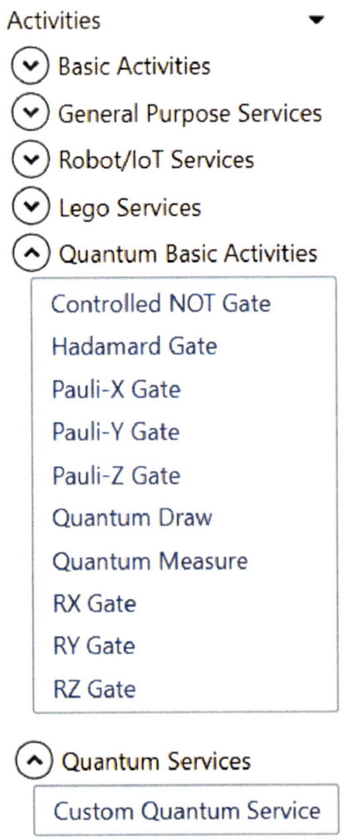

Fig. 12. VIPLE built-in Quantum Basic Activities and Quantum Services

5 Conclusions and Future Work

This paper presented the development of visual programming languages in business and education applications. VIPLE, developed by the authors, was presented and its new functions were introduced. VIPLE has been used as a teaching tool in many schools and universities around the world. The basic parts can be used at different grade levels in schools and the advanced parts, such as traffic simulation, autonomous driving, and quantum computing are being used at senior and graduate levels at universities.

Acknowledgments. This research is supported by the general fund of the IoT and Robotics Education Lab at Arizona State University.

References

1. Chen, Y., De Luca, G.: Service-Oriented Computing and System Integration: Software, IoT, Big Data, and AI as Services, 8th edn. Kendall Hunt Publishing (2022)

2. Megargel, A., Poskitt, C., Shankararaman, V.: Microservices orchestration vs. choreography: a decision framework. In: IEEE 25th International Enterprise Distributed Object Computing Conference (EDOC) (2021). https://doi.org/10.1109/EDOC52215.2021.00024
3. De Luca, G., Li, Z., Mian, S., Chen, Y.: Visual programming language environment for different IoT and robotics platforms in computer science education. CAAI Trans. Intell. Technol. **3**(2), 119–130 (2018). https://doi.org/10.1049/trit.2018.0016
4. Chen, Y., De Luca, G.: VIPLE: visual IoT/robotics programming language environment for computer science education. In: 2016 IEEE International Parallel and Distributed Processing Symposium Workshops (IPDPSW) (2016). https://doi.org/10.1109/ipdpsw.2016.55
5. De Luca, G., Chen, Y.: Semantic analysis of concurrent computing in decentralized IoT and robotics applications. In: Proceedings of ISADS, Utrecht, the Netherlands, pp. 95–102 (2019)
6. Boshernitsan, M., Downes, M.: Visual Programming Languages: A Survey, University of California, Berkeley, Technical Report No. UCB/CSD-04–1368, December 2004. http://www2.eecs.berkeley.edu/Pubs/TechRpts/2004/CSD-04-1368.pdf
7. Ray, P.P.: A survey on visual programming languages in internet of things. Sci. Program. (2017).https://doi.org/10.1155/2017/1231430
8. Tsai, W.T., Paul, R., Xiao, B., Cao, Z., Chen, Y.: PSML-S: a process specification and modeling language for service oriented computing. In: The 9th IASTED International Conference on Software Engineering and Applications (SEA), Phoenix, pp. 160–167 (2005)
9. BPMN, Business Process Model and Notation 2.0 (2010). https://www.omg.org/spec/BPMN/2.0/
10. Scribner, K.: Microsoft Windows Workflow Foundation Step By Step. Microsoft Press (2007)
11. Saraswathi, R., Singh, J.: Oracle SOA BPEL Process Manager 11gR1 - A Hands-on Tutorial. Packt (2013)
12. AWF, Amazon Simple Workflow Service, Developer Guide, API Version 2012–01–25. https://docs.aws.amazon.com/amazonswf/latest/developerguide/swf-dg.pdf#swf-dg-intro-to-swf
13. Getting Started with Scratch (2022). https://sip.scratch.mit.edu/wp-content/uploads/2020/06/Getting-Started-With-Scratch-3.0.pdf
14. Rogler, H.: Alice Programming, 2nd edn. Kendall Hunt Publishing (2016)
15. Lang, K., Tezel, S.: Become an App Inventor. MITeen Press (2022)
16. VPL, Microsoft Visual Programming Language. https://en.wikipedia.org/wiki/Microsoft_Visual_Programming_Language
17. Jawaharlal, M.: LEGO EV3 Robotics: A Guide for Educators. Red Gerbera (2015)
18. TORCS, the open race car simulator, Sourceforce. https://sourceforge.net/projects/torcs/
19. De Luca, G., Chen, Y.: Visual IoT/robotics programming language in pi-calculus. In: 2017 IEEE 13th International Symposium on Autonomous Decentralized System (ISADS) (2017). https://doi.org/10.1109/ISADS.2017.32
20. Zhao, S., Li, Y., Wang, Y., Chen, Y.: Validating trustworthy service composition through VIPLE and pi-calculus. Int. J. Simul. Process Model. **15**(1–2), 76–88 (2020). https://doi.org/10.1504/IJSPM.2020.106971
21. Zhang, Z., De Luca, G., Archambault, B., Chavez, J., Rice, B.: Traffic dataset for dynamic routing algorithm in traffic simulation. J. Artif. Intell. Technol. (2022). https://doi.org/10.37965/jait.2022.0106
22. De Luca, G.: A survey of NISQ era hybrid quantum-classical machine learning research. J. Artif. Intell. Technol. **2**(1), 9–15 (2021). https://doi.org/10.37965/jait.2021.12002
23. Zhang, Z., De Luca, G., Chen, Y.: Dynamic Dijkstra algorithm traffic simulation (2022). https://venus.sod.asu.edu/VIPLE/Videos/DynamicDijkstra.mp4

Pedestrian Detection Model Algorithm Optimization Based on Deep Convolutional Neural Network

Yizeng Wang[1,2], Hu Hao[1], Xiaoqing Zeng[3], and Dongliang Feng[4](✉)

[1] State Key Laboratory of Ocean Engineering, Shanghai Jiao Tong University, Shanghai 200240, China
{wangyizeng,hhu}@sjtu.edu.cn
[2] Department of Computer Science, City University of Hong Kong, Hong Kong 999077, China
[3] The Key Laboratory of Road and Traffic Engineering, Ministry of Education, Tongji University, No. 4800 Cao'an Road, Shanghai 201804, China
zengxq@tongji.edu.cn
[4] Shanghai Municipal Engineering Construction and Development Co., Ltd., Shanghai, China
fdlgzkm2007@126.com

Abstract. Pedestrian detection is a classical problem in computer vision and has been a difficult problem to solve for a long time. Compared with face detection, it is very difficult to accurately detect pedestrians in various scenarios because of the complex posture of human body, larger deformation, and more serious problems such as attachment and occlusion. This paper focuses on the typical pedestrian detection model - YOLO model. Through experiments, the principle of pedestrian detection model algorithm and its model effect are studied to solve the difficulties in pedestrian detection. In YOLO, logistic regression is used to predict the object score of each boundary box, and multi-scale fusion is used to make prediction. By observing the mAP index, it is concluded that the YOLO algorithm has a good effect on single-label pedestrian detection, and the calculation efficiency is high.

Keywords: Pedestrian Detection · YOLO · Target Detection · Darknet

1 Introduction

Pedestrian detection is a classical problem in computer vision, and it has been difficult to solve for a long time. Compared with face detection, it is more difficult to accurately detect pedestrians in various scenarios due to the complexity of human body posture, larger deformation, more serious problems such as attachments and occlusion. The problems to be solved by pedestrian detection are as follows: it is necessary to find out all the pedestrians in the image or video frame, including their positions and sizes, which are usually represented by rectangular boxes; Similar to face detection, this is a typical target detection problem. Pedestrian detection technology has a very strong practical value, can be combined with pedestrian tracking, pedestrian recognition and other technologies, applied to automobile unmanned driving system, intelligent robot, intelligent

X. Zeng et al. (Eds.): ITASC 2022, LNEE 1042, pp. 14–24, 2023.
https://doi.org/10.1007/978-981-99-2252-9_2

video monitoring, human behavior analysis, passenger flow statistics system, intelligent transportation and other fields. Due to the flexibility of human body, there will be a variety of posture and shape, its appearance is greatly affected by dress, posture, perspective and other factors, but also facing the impact of occlusion, lighting and other factors, which makes pedestrian detection become a very challenging topic in the field of computer vision.

In order to solve the above challenges, scholars worldwide put forward a large number of methods in recent years and achieved many results. Early algorithms used some simple methods in image processing and pattern recognition, but the accuracy was low. With the increase of training sample size, such as INRIA database, Caltech database and TUD pedestrian database, the accuracy of the algorithm is increasing. A landmark achievement in the field of pedestrian detection is the pedestrian detection algorithm based on HOG + SVM proposed by Dalal and Triggs in CVPR 2005 [1]. HOG (Histogram of gradient direction) is based on edge feature, which utilizes edge orientation and intensity information. Firstly, the gradient is calculated by using images of fixed size, and then the grid is divided. The gradient orientation and intensity at each point are calculated to form the histogram of the gradient direction distribution of all pixels in the grid. Finally, the histogram features are summarized to form the whole histogram. After obtaining the HOG feature of the candidate region, it is necessary to classify the region by using the classifier Class to determine whether it is a pedestrian or background area. Dalal [2] proposed a pedestrian detection scheme using cascaded convolutional network. Angelova [3] proposed a method called SA-FastRCNN, and it improved Fast R-CNN based on the characteristics of pedestrian detection. Since the features extracted from large-size pedestrians and small-size pedestrians showed significant differences, the author designed two subnetworks for large-size pedestrians and small-size pedestrians respectively for detection. RepLoss [4] was proposed by face + + and its main goal was to solve the occlusion problem. In pedestrian detection, there may be a problem of occlusion in dense human body detection. Object occlusion can be divided into two types: in-class occlusion and inter-class occlusion. In-class occlusion refers to mutual occlusion between similar objects. In pedestrian detection, such occlusion accounts for a larger proportion and seriously affects the performance of pedestrian detector. To solve this problem, the authors designed a loss function called RepLoss [authors of this paper or from a reference? Wang [5] proposed a pedestrian detection algorithm called Hyper-Learner, which improved from faster R-CNN. In this article, the authors analyze the difficulties of pedestrian detection: The distinction between pedestrian and background is low, and it is very difficult to accurately define a single pedestrian in crowded scenes. The authors introduce some additional features to solve these problems. In order to send these additional features into the convolutional network for processing, the author adds a branch network on the basis of VGG network. Together with the characteristics of the main network, it is fed into the RPN for processing.

YOLO had gradually become the mainstream method for pedestrian detection in machine learning. A new object detection algorithm, R-CNN, was introduced in the paper and mean average precision (mAP) was improved by over 30% relative to the previous best result on VOC 2012 by combining region proposals with high-capacity convolutional neural networks (CNNs) and utilizing supervised pre-training for an auxiliary task followed by domain-specific fine-tuning when labeled training data was scarce [7]. A fully convolutional network, the Region Proposal Network (RPN), was presented and the sharing of full-image convolutional features with the detection network by the RPN led to cost-free region proposals and improved object detection performance, resulting in state-of-the-art object detection accuracy on multiple datasets with a frame rate of 5 fps [8]. YOLO was a real-time object detection approach using a single, optimized neural network that predicted bounding boxes and class probabilities, outperforming existing methods in generalization from natural images to other domains [9]. And even further, YOLO9000 was a state-of-the-art real-time object detection system capable of detecting over 9000 object categories with improved accuracy and speed, achieved through joint training on object detection and classification data [10]. The paper presented a method for improved pedestrian detection in transformer substations using a combination of Gaussian Mixture Model and YOLO, achieving a 20% improvement over single method results for increased safety near high voltage. A method for improved pedestrian detection in transformer substations through the combination of Gaussian Mixture Model and YOLO was carried out and 20% improvement over single method results for increased safety near high voltage was achieved [11]. In addition, the proposed MAF-YOLO method used a multi-modal feature extraction module and a modal weighted fusion module to achieve real-time and precise pedestrian detection at nighttime through the utilization of a dual attention module and improved loss function and anchor box size [12].The rest of the paper is organized as follows. Section 2 generated the research scope of pedestrian detection based on YOLOv3, Sect. 3 introduced the methodology of the model and the process of model selection Sect. 4 showed the empirical experiment include of evaluation indicators, experimental procedure and result analysis, and the conclusion had been gven in the Sect. 5.

2 Research Scope

The YOLO model of pedestrian detection is studied mainly. Through practical experiments, the principle of YOLOv3 [6] algorithm and its model effect are studied to solve the difficulties in pedestrian detection.

The main difficulties to be solved in pedestrian detection include: first, the appearance difference is large, including perspective, posture, clothing and attachments, lighting, imaging distance, etc.; Pedestrians look very different from different angles; As pedestrians in different poses wear different clothes, as well as the shadow of umbrella, hat, scarf, luggage and other attachments, the appearance of the human body at a long distance and a close distance is very different. The second is the problem of occlusion. In many application scenarios, pedestrians are very dense, there are serious occlusion, only part of the human body can be seen, which is conducive to detection. Algorithms pose serious challenges. Third, the background is complex. Whether indoor or outdoor, the background for pedestrian detection is usually complicated. The appearance, shape, color and texture of some objects are very similar to human bodies, so the algorithm cannot accurately distinguish them. The fourth is the detection speed. Pedestrian detection generally adopts a complex model with a large amount of computation. It is difficult to achieve real-time detection and generally requires a lot of optimizations.

3 Pedestrian Detection Model

Nowadays, object detection based on deep learning has gradually become the core technology in the fields of autonomous driving, video surveillance, machining, intelligent robots and so on. Most of the existing target detection algorithms have relatively high accuracy, but their speed is slow and cannot adapt to the real-time demand of target detection in the industry. YOLO algorithm has won unanimous praise for its speed and accuracy. So far, the YOLO algorithm has gone through five iterations and has achieved great improvement in speed and accuracy.

3.1 Model Introduction

The core idea of YOLOv1 is to solve target detection as a regression problem. Instead of training the network by sliding window or proposal extraction, YOLO directly selects the whole graph training model. The advantage of this is that you can better distinguish between the target and the background area. YOLOv1 will first shrink the original image to 448 × 448 inches for easy division later. However, the image is divided into S × S regions. Note that the concept of this region is different from that of dividing the image into N regions and throwing it into the algorithm as mentioned above. The region mentioned above is to crop the picture, or to say, to input a local pixel of the picture into the algorithm, while the division region is logical.

If the center of an object falls on a cell, the cell is responsible for predicting the object. Each cell needs to pre-test B bbox values and predict a confidence scores for each bbox value. Then, the predictive analysis is carried out in each cell. The confidence is not only the probability that the boundary box is the object to be detected, but the probability that the boundary box is the object to be detected multiplied by the IoU of the boundary box and the real position (the intersection between the frames divided by the union).By multiplying the intersection ratio, the accuracy of position prediction of the boundary box is reflected:

$$confidence = P(Object) \times IoU_{pred}^{truth} \tag{1}$$

Each bounding box corresponds to five outputs, which are x, y, w, h and confidence. Where x and y represent the deviation of the center of the boundary box from the boundary of the grid cell it is in. w and h represent the proportion of the true width and height of the boundary box to the whole image. x, y, w, and h have been restricted to the interval [0,1]. In addition, each cell generates C conditional probabilities (Fig. 1).

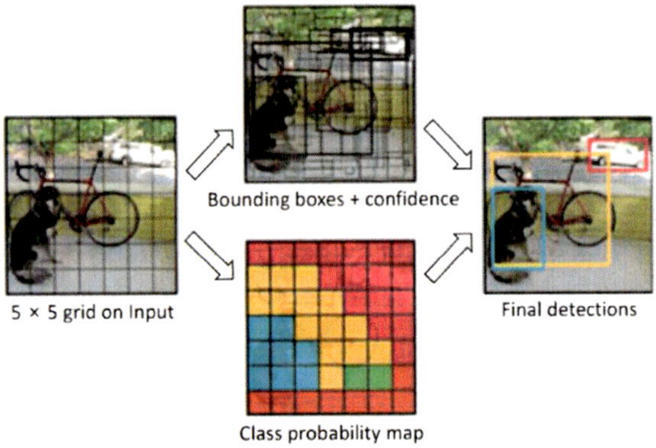

Fig. 1. YOLO Prediction Graph

In the non-maximum suppression stage of test, for each boundary box, Formula 2 is used to measure whether the box should be retained.

$$confidence \times P(Class_i|Object) = P(Class_i) \times IoU_{pred}^{truth} \tag{2}$$

This is the individual classification confidence score of each cell, which contains both the predicted category information and the accuracy of the bbox value. A threshold is set to filter out the low class-specific confidence scores, and leave the rest for non-extremely large value suppression to obtain the final calibration frame.

The YOLO web design follows, but differs from, GoogleNet's ideas. YOLO uses 24 cascaded convolution layers and 2 fully connected layers, among which CONV layer includes 3*3 and 1*1 Kernel. The last FC layer is the output of YOLO network, with a length of S*S*(B*5 + C) = 7*7*30. In addition, the simplified version of YOLO - Small network includes 9 cascaded CONV layers and 2 FC layers. Since the number of CONV layers is much less, the speed of YOLO - Small is much faster than that of YOLO. YOLO network architecture is shown in Fig. 2.

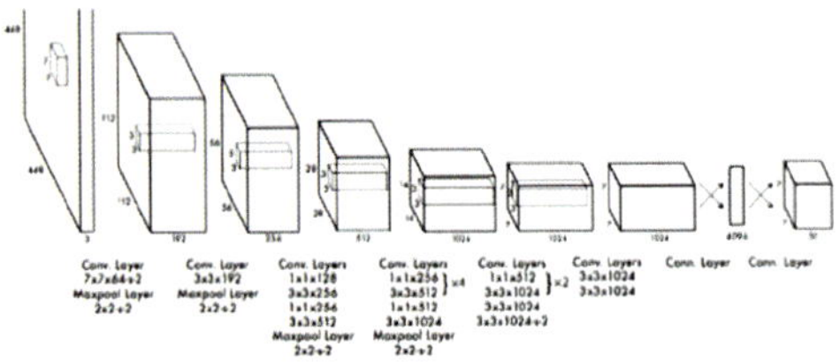

Fig. 2. YOLO Network Architecture

All of YOLOv1 methods use mean square error as loss function. It consists of three parts: coordinate error, IOU error and classification error. Considering the contribution rate of each loss, YOLOv1 sets the weight λcoord = 5 for the coordinate error. When calculating IoU error, the contribution value of IoU error to network Loss is different between the grid containing objects and the grid without objects. If the same weight is used, the confidence value of the cell that does not contain objects is approximately 0, which is amplified in a disguised way The influence of the confidence error of the cell containing the object on the calculation of the network parameter ladder. To solve this problem, YOLO uses λ noobj = 0.5 to correct (confidence error) iouErr. For equal error values, the influence of large object error on detection should be less than that of small object error. This is because the proportion of the same positional deviation for large objects is much smaller than the proportion of the same deviation for small objects. YOLO improves this problem by taking the square root of the information items of object size (w and h), but it cannot completely solve this problem.

YOLOv1 uses the standard linear activation function for the last layer, while the leaky Rectified linear activation function is used for the other layers.

3.2 Model Selection

In the process of the development of YOLO model, YOLOv2 made a series of improvements on the basis of YOLOv1 and reached state of the Art at the same time. YOLOv2 can adapt to different input sizes and adjust detection accuracy and speed as required. However, YOLOv2 still cannot provide a good solution for overlapping classification.

YOLOv3 is the target detection network with the most balanced speed and accuracy. Through the integration of a variety of advanced methods, YOLO series of short board complete. After YOLO9000 [7], the system uses dimensional clustering as anchor boxes to predict boundary boxes, and the network predicts 4 coordinates for each boundary box. Logistic regression is used in YOLOv3 to predict the object score for each bounding box. If the previous bounding box overlaps the Ground Truth object more than any other bounding box before it, the value should be 1. If the previous boundary box is not the best but does overlap the ground truth object above a certain threshold value, we will ignore this prediction and use 0.5 as the threshold value. Unlike YOLOv2, our system only assigns a bounding box to each Ground truth object. If the previous bounding box is not assigned to the Grounding Box object, there is no loss to the coordinates or category prediction.

In YOLOv3, each box uses a multi-label classification to predict which classes the bounding box might contain. This algorithm does not use softmax, YOLOv3 uses a separate logical classifier. In the training process, binary cross entropy loss is used for category prediction. For overlapping labels, the multi-label approach can better simulate the data.

YOLOv3 uses multi-scale fusion to make predictions. The original YOLOv2 has one layer: Passthrough Layer, assuming that the size of the feature map extracted last is 13*13, Therefore, the function of this layer is to connect the feature map of 26*26 of the previous layer with the feature map of 13*13 of this layer. Such operation is also to strengthen the accuracy of YOLO algorithm for small target detection. This idea is further enhanced in YOLOv3, which uses a similar upsampling and fusion approach to FPN, in multiple applications. According to the detection on the feature map of scale, the detection effect of small targets is significantly improved. Although YOLOv3 predicts 3 boundary boxes per grid cell, which seems less than YOLOv2 predicts 5 boundary boxes per grid cell, because YOLOv3 adopts feature fusion of multiple scales, the number of boundary boxes is larger than before.

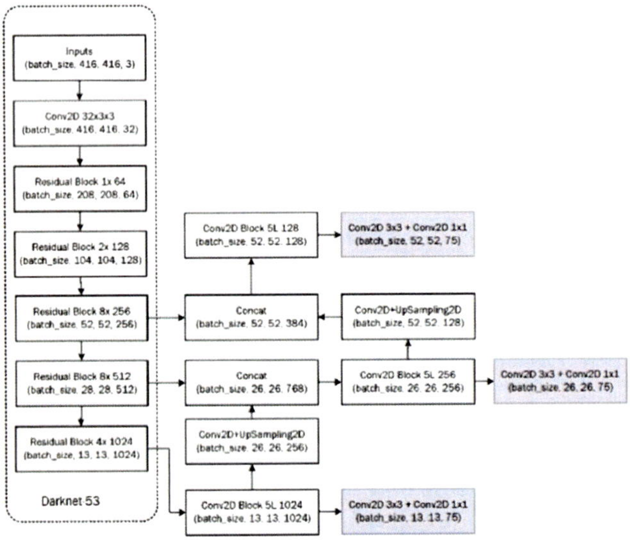

Fig. 3. Darknet-53 network architecture

YOLOv3 uses Darknet-53 network to achieve feature extraction. Darknet-53 network architecture is shown in Fig. 3. Adding a hybrid approach to residual networks in Darknet-19, using continuous 3 × 3 and 1 × 1 convolutional layers, but now with some shortcut connections, YOLOv3 expanded it to 53 layers and called it Darknet-53.

4 Empirical Experiment

This section begins with describing the data sets used to train and test the model, as well as the evaluation metrics. Then the training process of the model is introduced, and the results of the model are analyzed.

The data set used in this experiment is coco data set, which is a large and rich object detection, segmentation and subtitle data set. This data set takes scene understanding as the target, which is mainly extracted from complex daily scenes, and targets in images are calibrated through accurate segmentation. Images include 91 categories of objects, 328,000 images, and 2,500,000 labels. The data used in this experiment is part of coco2014 segmentation, and the test set is 3000 pictures.

4.1 Evaluation Indicators

In order to evaluate the model, mAP index was used in this experiment. MAP refers to Mean Average Precision or Average AP value, which is calculated for the Average AP value of multiple verification sets and used as an indicator to measure the detection accuracy in object detection. MAP is also suitable for multi-label classification task and multi-label image classification task, but in this experiment, it is only single-label classification task.

4.2 Experimental Procedure

The model uses dimensional clustering as anchor boxcs to predict edge boxes. During training, the sum of squared error losses is used. The number of objects in each bounding box is predicted using logistic regression. This value should be 1 if the ground Truth object was reduplicated by the previous bounding box than any other bounding box. If the previous boundary box is not the best, but the ground truth object does overlap above a certain threshold, this prediction will be ignored. The threshold is set to 0.5. The model only assigns a boundary box to each ground truth object. If the previous bounding box is not assigned to the Grounding Box pair, there is no loss to the coordinates or category prediction. Each box uses a multi-label classification to predict which classes the bounding box might contain. Instead of using softmax, the model uses a separate logical classifier. In the training process, the model uses binary cross entropy loss for category prediction.

YOLOv3 predicts three boxes at different scales. Type uses similar concepts to extract features of these scales to form a pyramid network. From the basic feature extractor, several layers of codeling are added. In the COCO dataset, 3 boxes are predicted for each scale, so for 4 bounding box offsets, 1 target prediction and 80 category prediction, the tensor is N × N × [3 * (4 + 1 + 80)]. Next, the feature maps were obtained from the previous two layers and upsampled by 2 times. Feature maps were taken from earlier versions of the network and merged with upsampled features using element-wise addition. Several convolution layers are added to process the combined feature graph, and a similar tensor is eventually predicted. Perform the same design again to anticipate the final scale of the box.

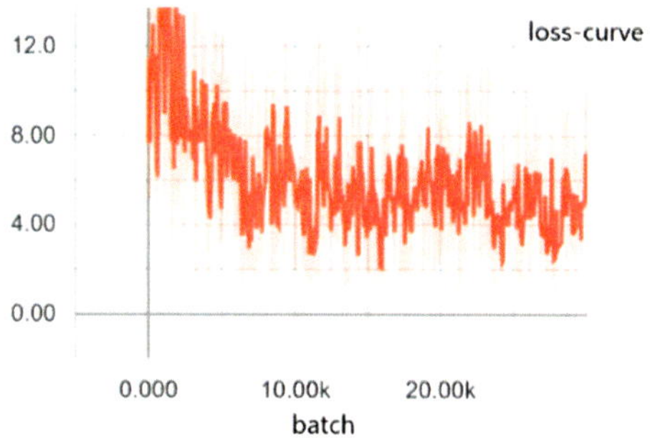

Fig. 4. Loss Curve of Boundary Box

Use k-means clustering to determine the priors of the boundary boxes. Nine clusters and three scales were selected, and then the clusters were evenly divided on the whole scale. The 9 clusters are: (10 × 13); (16 × 30); (33 × 23); (30 × 61); (62 × 45); (59 × 119); (116 × 90); (156 × 198); (373 × 326). Use Darknet-53 network for training and testing. Figure 4 is the Loss curve of the boundary box. It can be seen that the Loss value has converged.

Figure 5 shows the Loss curve of the model, and the loss value has converged.

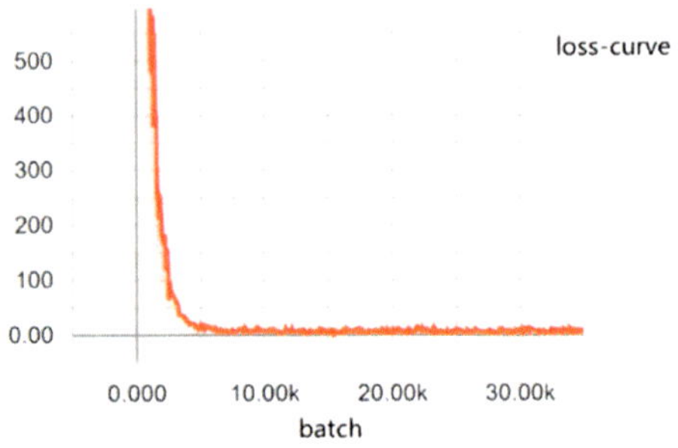

Fig. 5. Loss Curve of YOLOv3 Model

4.3 Result Analysis

The mAP of the model is shown in Fig. 6. The average AP value of multiple validation sets is 72.32%.

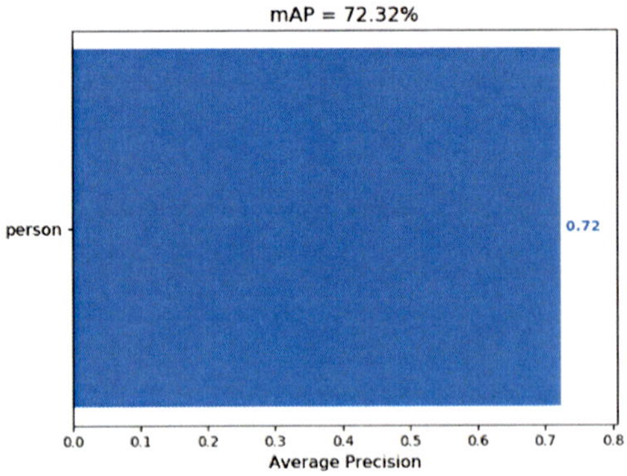

Fig. 6. mAP Values

The P-R curve of the model is shown in Fig. 7. P-R curve is a two-dimensional curve with precision and recall as the vertical and horizontal coordinates. The overall trend is drawn by selecting the accuracy and recall rates corresponding to different thresholds. The higher the accuracy, the lower the recall. The area enclosed by the P-R curve is the AP value. The classifier works better from the diagram.

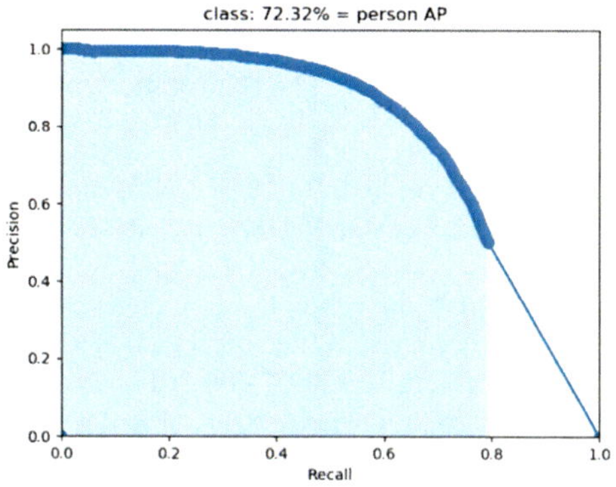

Fig. 7. P-R Curve

5 Conclusion

Pedestrian detection experiments of YOLOv3 model were carried out on COCO dataset. By observing mAP index, YOLOv3 was a good detection model. Fast speed, high accuracy. The mAP on the COCO dataset was about 73.32%. However, in this experiment, only pedestrians were detected, and it was a single classification label detection, so mAP performed well. In the multi-classification label detection, the effect of the model should be analyzed through other experiments. Pedestrian detection is an interesting and difficult subject, with broad research prospects and great practical significance. There are many applications in many practical scenarios and events, and it is worth further exploration.

Acknowledgments. The project is supported by Science and Technology Commission of Shanghai Municipality (Number 22dz1208505, 20DZ1202900, 19DZ1204200).

References

1. Mao, J., Xiao, T., Jiang, Y., Cao, Z.: What can help pedestrian detection? In: Proceedings of the IEEE Conference on Computer Vision & Pattern Recognition. IEEE (2017)
2. Dalal, N.: Histograms of oriented gradients for human detection. In: Proceedings of CVPR (2005)
3. Angelova, A., Krizhevsky, A., Vanhoucke, V., et al.: Real-time pedestrian detection with deep network cascades. In: British Machine Vision Conference (2015)
4. Li, J., Liang, X., Shen, S.M., et al.: Scale-aware fast R-CNN for pedestrian detection. IEEE Trans. Multimedia **99**, 1–1 (2015)
5. Wang, X., Xiao, T., Jiang, Y., et al.: Repulsion loss: detecting pedestrians in a crowd. In: Proceedings of the 2018 IEEE/CVF Conference on Computer Vision and Pattern Recognition (CVPR). IEEE (2018)
6. Redmon, J., Farhadi, A.: YOLOv3: an incremental improvement. arXiv e-prints, 2018
7. Girshick, T., Donahue, J., Darrell, T., Malik, J.: Rich feature hierarchies for accurate object detection and semantic segmentation. In: Proceedings of the 2014 IEEE Conference on Computer Vision and Pattern Recognition, Columbus, OH, USA, pp. 580–587 (2014). https://doi.org/10.1109/CVPR.2014.81

8. Ren, U., He, K., Girshick, R., Sun, J.: Faster R-CNN: towards real-time object detection with region proposal networks. IEEE Trans. Pattern Anal. Mach. Intell. **39**(6), 1137–1149 (2017). https://doi.org/10.1109/TPAMI.2016.2577031
9. Redmon, L., Divvala, S., Girshick, R., Farhadi, A.: You only look once: unified, real-time object detection. In: Proceedings of the 2016 IEEE Conference on Computer Vision and Pattern Recognition (CVPR), Las Vegas, NV, USA, pp. 779–788 (2016). https://doi.org/10.1109/CVPR.2016.91
10. Redmon, M., Farhadi, A.: YOLO9000: better, faster, stronger. In: Proceedings of the 2017 IEEE Conference on Computer Vision and Pattern Recognition (CVPR), Honolulu, HI, USA, pp. 6517–6525 (2017). https://doi.org/10.1109/CVPR.2017.690
11. Peng, R., et al.: Pedestrian detection for transformer substation based on gaussian mixture model and YOLO. In: 2016 8th International Conference on Intelligent Human-Machine Systems and Cybernetics (IHMSC), Hangzhou, China, pp. 562–565 (2016). https://doi.org/10.1109/IHMSC.2016.130
12. Xue, Y., Ju, Z., Li, Y., Zhang, W.: MAF-YOLO: multi-modal attention fusion based YOLO for pedestrian detection. Infrared Phys. Technol. **118**, 103906 (2021) ISSN 1350–4495. https://doi.org/10.1016/j.infrared.2021.103906

Study on Evaluation Method of Ecological Road Based on Vague - Extenics

Xiaoqing Zeng[1], Yuan Liu[1], Yang Lu[2](✉), Junxiang He[1], Yizeng Wang[3,4], and Qiao He[5]

[1] The Key Laboratory of Road and Traffic Engineering, Ministry of Education, Tongji University, No.4800 Cao'san Road, Shanghai 201804, China
zengxq@tongji.edu.cn
[2] Shanghai Municipal Engineering Construction and Development Co., Ltd., Shanghai, China
Luli308501@163.com
[3] State Key Laboratory of Ocean Engineering, Shanghai Jiao Tong University, Shanghai 200240, China
wangyizeng@sjtu.edu.cn
[4] Department of Computer Science, City University of Hong Kong, Hong Kong 999077, China
[5] School of Information, University of Technology of Belfort-Montbéliard, 90000 Belfort, France
qiao.he@utbm.fr

Abstract. In view of the lack of a relatively scientific and complete evaluation system in the field of urban ecological road construction, the traditional evaluation methods cannot meet the needs of quantitative and single-object overall evaluation of ecological roads. In this paper, an ecological road evaluation method based on Vague compound extensibility is proposed. This study established a comprehensive evaluation index system for ecological roads and determined the index weights based on the free entropy weight structure method. Through the case experiment analysis, the evaluation index standard and the object to be evaluated are integrated into the Vague, so as to calculate the similarity of each index and at each level of the object to be evaluated. The comprehensive evaluation level of the ecological road in the evaluation section is determined as the second level. The ecological road evaluation method of Vague-Extensology can still comprehensively deal with different types of indicators in the case of relatively single data and achieve good results in qualitative and quantitative evaluation.

Keywords: Ecological road · Vague Complex Extenics · Structural Entropy Weight Method · Evaluation index system

1 Introduction

In recent years, countries around the world have been vigorously advocating the construction of global green ecological civilization. The United States promulgated the American Recovery and Reinvestment Act in 2009 [1]. The development direction of ecological environment protection and transportation investment is emphasized. In response to

X. Zeng et al. (Eds.): ITASC 2022, LNEE 1042, pp. 25–38, 2023.
https://doi.org/10.1007/978-981-99-2252-9_3

the UN's initiative on green ecological development and global sustainable transportation, France promulgated the "Law on Future Transportation Orientation" in 2019 [2]. It is pointed out that the transportation construction should speed up the ecological transformation and actively participate in the ecological environment protection. The Implementation Plan for Promoting Ecological Progress in Transport issued by China in 2017 [3]. Proposed to actively promote the construction of ecological civilization in the field of transportation. The construction of ecological civilization in the field of transportation has become an important national development strategy of all countries in the world. In the process of promoting this strategy, the construction of ecological road is an extremely important link.

In many countries, the concept of ecological protection has been deeply integrated into road transport construction [4]. Ecological road is to use advanced technology in the whole life cycle process of road design, construction and operation, save resources, protect the environment, and form a new road traffic system with safe and comfortable driving and harmonious ecological landscape. In order to promote the application of green ecological concept in the field of transportation, many related institutions and scholars have put forward different evaluation systems and evaluation standards [5]. In order to strengthen the green sustainability of road facilities construction, the study of ecological road evaluation method has become one of the important research directions in the field of green transportation. Based on Vague - extenics, this paper analyzes the advantages and disadvantages of several evaluation methods and their applicable scope, and proposes a new ecological road evaluation method based on vague - extenics.

The rest of the paper is organized as follows. Section 2 analyses the demand of ecological road evaluation methods. Section 3 proposed an ecological road evaluation method based on Vague- Extenics. Section 4 conducts a case study on an expressway and gives the evaluation grade (Fig. 1).

Fig. 1. Evaluation site selection of Ecological Road

2 Demand Analysis of Ecological Road Evaluation Methods

At present, there are many related evaluation methods for ecological roads. Among them, Analytic Hierarchy Process (AHP) [6] is a method that combines qualitative and quantitative analysis, but the evaluation results are greatly influenced by the subjective will of review experts, and are mostly used for program selection and weight coefficient calculation; the extension matter-element evaluation method [7] can deal with the variability index, but the data processing is more complicated, and it is used to deal with the incompatibility of each index; the Fuzzy Synthetical Evaluation method [8] can make the evaluation The ambiguity of standards and influencing factors is reflected, but the accuracy of evaluation results is unstable, and it is used for the evaluation of indicators with strong subjective factors. Therefore, in order to meet the conditions of the evaluation methods required by the ecological road, the advantages and disadvantages of different evaluation methods and the scope of application are compared and used in combination.

According to the characteristics of ecological road evaluation indicators, the proposed ecological road evaluation method must meet the following requirements: (1) The evaluation method must be compatible with both quantitative and qualitative index evaluation requirements. (2) Under the circumstance that the evaluation data is difficult to obtain and the data volume is small, the ecological road grade evaluation for a single evaluation object can still be completed well. (3) In the face of the low correlation in all aspects of the ecological road system, the evaluation method can integrate the indicators of various evaluation standards to give an overall evaluation of the evaluation object.

3 Ecological Road Evaluation Method Based on Vague - Extenics

3.1 Selection of Ecological Road Evaluation Index

The ecological road evaluation index system was obtained and established by consulting the engineering design book and expert scoring method, as shown in Table 1.

Table 1. Ecological road comprehensive evaluation index system

classification	Serial number	indicators	classification	Serial number	indicators
Conservation of resources	A1	Land resource occupation	Construction of quality	C1	Long life pavement
	A2	Rationality of road space		C2	Functional road surface
	A3	Drainage and storage engineering		C3	Linear design continuity

(continued)

Table 1. (*continued*)

classification	Serial number	indicators	classification	Serial number	indicators
	A4	Recyclable material utilization		C4	Quality of bridge and tunnel
	A5	Energy saving measures		C5	Construction of standardization Organization and Management
	A6	Renewable energy Sources Energy saving management mechanism		C6	Information management
	A7	Energy saving management mechanism		C7	Application of BIM technology
Ecological protection	B1	Social environmental protection		C8	Preventive maintenance
	B2	Biological habitat conservation	Smart Management	D1	Intelligent transportation system applications
	B3	Ecological continuity protection		D2	Of safety facilities Layout and maintenance
	B4	Protection of vegetation		D3	Traffic Congestion index
	B5	Water body protection		D4	Traffic Safety Index
	B6	Air pollution protection		D5	Traffic Management System
	B7	Acoustic pollution control		D6	Transportation investment coordination coefficient
	B8	Light pollution control	Serve the humanities	E1	Public satisfaction

(*continued*)

Table 1. (*continued*)

classification	Serial number	indicators	classification	Serial number	indicators
	B9	Dangerous goods management		E2	Landscape Design
	B10	Heat island effect control		E3	Publicity and promotion
	B11	Environmental management systems and funding		E4	Eco-environmental Training

3.2 Index Weight Calculation Based on the Theory of Free Entropy Weight Structure

At present, there are many commonly used weight coefficient calculation methods. Considering the advantages and disadvantages of the main and objective weight methods, this paper uses the free structure entropy weight method to calculate the weight coefficient. Delphi method is used to form the index fundamental importance matrix. Experts from related fields are selected and questionnaires are issued. The index fundamental importance matrix P_{mn} is obtained by sorting out the survey results in an anonymous form, which m represents the number of indicators and n the number of experts. x_{ij} represents the importance score of expert j on the i indicator, which x_{ij} is 0, 1, 2, 3 and 4. In order to form the entropy importance matrix, it is necessary to calculate the entropy membership. Firstly, the information entropy needs to be calculated:

$$H_i = -k\sum\nolimits_{j=1}^{n} f_{ij} \ln f_{ij} \tag{1}$$

The above equation represents the information entropy value of index i in the presence of a kind of importance score j, where k is a constant. Then the entropy of each case is:

$$h_{ij} = -kf_{ij} \ln f_{ij} \tag{2}$$

$f_{ij} = \frac{\varepsilon - x_{ij}}{\varepsilon - \theta}$, θ is constant; $\varepsilon = \theta + 2k = \frac{1}{\ln(\varepsilon - \theta)}$, ε and k are all constants.

μ_{ij} is set as the membership function of entropy value of x_{ij}.

$$\mu_{ij} = \frac{\ln(\varepsilon - \theta)}{\ln\left(\varepsilon - x_{ij}\right)} \tag{3}$$

$\left\{x_{ij} | x_{ij} = 0, 1, 2, \ldots, 6\right\}$; $\theta = x_{ij}^{max}$; $\varepsilon = x_{ij}^{max} + 2$.$\mu_{ij}$ is a monotonically increasing function of x_{ij}. The larger the μ_{ij} is, the more orderly the indicator is, the more unified the score of the indicator is by experts, the higher the importance is, and the greater the weight assigned to the corresponding indicator should be.μ_{ij} Then the entropy importance matrix $S_{mn} = \left\{\mu_{ij}\right\}_{m \times n}$ is calculated by entropy membership degree.

After eliminating the "importance deviation" caused by the limited expert knowledge, the importance of the comprehensive entropy is obtained. μ_{ij} represents the entropy importance of the expert *j* on the index *i*. Suppose that each expert has the same impact on the evaluation of each index. Therefore, the entropy importance $\overline{\mu_i}$ of the index *i* can be obtained by averaging the entropy importance of *n* experts on the index.

$$\overline{\mu_i} = \frac{(\mu_{i1} + \mu_{i2} + \ldots + \mu_{in})}{n} \tag{4}$$

Since the experts participating in the questionnaire came from different professional fields, the "importance bias" of the index caused by the cognitive gap was considered. In order to eliminate the influence of scores with large discrepancies, the "importance bias" B_i is set as follows:

$$B_i = \left| \frac{(max(\mu_{i1}, \mu_{i2}, \ldots, \mu_{in}) - \overline{\mu_i}) + (min(\mu_{i1}, \mu_{i2}, \ldots, \mu_{in}) - \overline{\mu_i})}{2} \right| \tag{5}$$

Finally, the importance degree v_i of comprehensive entropy with the "importance deviation" eliminated is obtained.

$$v_i = \overline{\mu_i} \cdot (1 - B_i) \tag{6}$$

(4) The comprehensive weight w_i of index importance can be obtained by normalized processing:

$$w_i = \frac{v_i}{\sum_{i=1}^{m} v_i} \tag{7}$$

According to the index weight calculation method based on the theory of free entropy weight structure, the weight coefficients of 36 indicators of the ecological road comprehensive evaluation system are obtained, as shown in Table 2.

Table 2. Weight coefficient of ecological road comprehensive evaluation index system

Serial number	Coefficient of weight	Serial number	Coefficient of weight	Serial number	Coefficient of weight	Serial number	Coefficient of weight
A1	0.0371	A2	0.0328	A3	0.0243	A4	0.0251
A5	0.0261	A6	0.0262	A7	0.0286	B1	0.0275
B2	0.0364	B3	0.0381	B4	0.0398	B5	0.0391
B6	0.0391	B7	0.0251	B8	0.0250	B9	0.0255
B10	0.0240	B11	0.0358	C1	0.0233	C2	0.0233
C3	0.0233	C4	0.0244	C5	0.0222	C6	0.0241

(continued)

Table 2. (*continued*)

Serial number	Cocfficient of weight	Serial number	Coefficient of weight	Serial number	Coefficient of weight	Serial number	Coefficient of weight
C7	0.0223	C8	0.0272	D1	0.0296	D2	0.0240
D3	0.0243	D4	0.0255	D5	0.0286	D6	0.0255
E1	0.0269	E2	0.0235	E3	0.0219	E4	0.0255

3.3 Evaluation Model of Vague Compound Extenics

Vague sets were proposed by W.L. au and J. Bouehrer [9, 10]. Through the information of true membership, false membership and hesitation, the degree of support and opposition can be expressed at the same time, and the fuzzy characteristics of the research object can be expressed in detail. Based on Vague- extension evaluation method for ecological road is proposed in this paper.

Firstly, qualitative values are Vague transformed. In this study, qualitative indicators are evenly divided into 5 levels. Vague variable set is constructed as $\mathrm{P} = \{\mathrm{x_j}\}$, 5 level variables are defined as $\mathrm{x_j} = [\mathrm{t_j}, 1 - \mathrm{f_j}], \mathrm{t_j}, \mathrm{f_j} \in [0, 1], \mathrm{j} = 1, 2, \ldots, 5$. The determinate of high grade evaluation is true membership t_j, while that of low grade evaluation is false membership $\mathrm{f_j}$. The level variable set of Vague value about qualitative index is constructed as $\mathrm{P} = \{\mathrm{x_j}\}$, as shown in Table 3.

Table 3. Qualitative index level variable set of Vague value

Grade of evaluation	Vague values	Degree of hesitation
1	[0.8, 1]	0.2
2	[0.6, 0.8]	0.2
3	[0.4, 0.6]	0.2
4	[0.2, 0.4]	0.2
5	[0, 0.2]	0.2

Then, the quantitative value of benefit index is transformed by Vague.

$$\begin{cases} t_{ij} = \frac{x_{ij} - x_i^{min}}{x_i^{max} - x_i^{min}} \\ f_{ij} = \frac{x_i^{max} - x_{ij}}{x_i^{max} - x_i^{min}} \end{cases}, (i = 1, 2, \ldots, m; j = 1, 2, \ldots, n) \tag{8}$$

Transform of Vague value of quantitative value of cost type index.

$$\begin{cases} t_{ij} = \frac{x_i^{max} - x_{ij}}{x_i^{max} - x_i^{min}} \\ f_{ij} = \frac{x_{ij} - x_i^{min}}{x_i^{max} - x_i^{min}} \end{cases}, (i = 1, 2, \ldots, m; j = 1, 2, \ldots, n) \tag{9}$$

x_{ij} represents the quantity value of j corresponding to characteristic i, x_i^{max} represents the maximum value of characteristic i of all, and x_i^{min} represents the minimum value of characteristic i of all.

Similarity is calculated by similarity measure of Vague value [11–13]. Set $U, x \in U$ as a non-empty set, V and W are two Vague sets on U. $\gamma_V(x)$ and $\gamma_W(x)$ are the Vague values of x about Vague set V and W. The formula for calculating the similarity between index i of item v to be evaluated and the corresponding standard of the evaluation level j of the index is as follows:

$$Z(v_i, w_{ij}) = 1 - \frac{\left|\left[t_{vi}(1+\sigma_{vi}) - f_{vi}(1+\sigma_{vi})\right] - \left[t_{wij}(1+\sigma_{wij}) - f_{wij}(1+\sigma_{wij})\right]\right|}{4} - \frac{\left|t_{vi}(1+\sigma_{vi}) - t_{wij}(1+\sigma_{wij})\right| + \left|f_{vi}(1+\sigma_{vi}) - f_{wij}(1+\sigma_{wij})\right|}{4} \tag{10}$$

$[t_{vi}, 1 - f_{vi}]$ is the index value of the index i of the item v to be evaluated, and $[t_{wij}, 1 - f_{wij}]$ is the index value of the corresponding standard of the evaluation level j of the index i. Then the formula for calculating the overall similarity between the whole thing to be evaluated and the fourth evaluation level j is as follows:

$$Z_j = \sum\nolimits_{i=1}^{m} w_i Z(v_i, w_{ij}), (j = 1, 2, \ldots, 5) \tag{11}$$

Z_j is the overall similarity between v to be evaluated and the evaluation level j. w_i is the weight of the index$i$. Meanwhile, if the similarity Z_j between v and the evaluation level j is the largest, then the evaluation of the item to be evaluated is the evaluation level j.

$$Z = max(Z_1, Z_2, \ldots, Z_j) \tag{12}$$

4 Case Study

This study is based on Vague - extenics ecological road evaluation method based on vague - Extenics, taking the north highway (construction highway ~ Panlong Highway) of Chongming Island, Shanghai, China as the evaluation object (Fig. 2).

4.1 Vague Set of Evaluation Index Criteria

Based on the determination of index weight coefficient, qualitative index and quantitative index are transformed into Vague set according to Vague set according to the evaluation content and evaluation grade division standard of each single index, in which the maximum x_i^{max} and minimum $\mathrm{x}_\mathrm{i}^{\mathrm{min}}$ values of vague set are taken respectively from five evaluation levels of corresponding index, and the evaluation grade standard of 36 evaluation indexes is transformed into vague set.

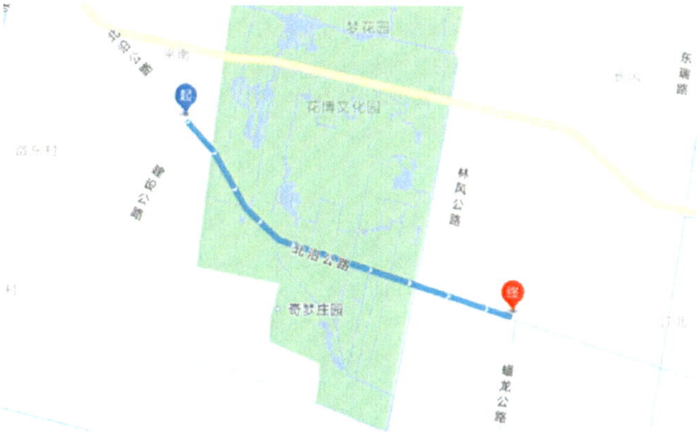

Fig. 2. Schematic diagram of North Extension Highway

Table 4. Index value Vague transformation results of evaluation objects

indicators	dimension	Index	Vague	indicators	dimension	Index	Vague
A1	%	10	[0.9, 0.9]	A2	m	5	[0.71, 0.71]
A3	%	77	[0.77, 0.77]	A4	%	75	[0.75, 0.75]
A5	qualitative	4	[0.8, 0.1]	A6	%	70	[0.7, 0.7]
A7	qualitative	L2 secondary	[0.6, 0.8]	B1	%	90	[0.9, 0.9]
B2	%	3	[0.7, 0.7]	B3	km	1.6	[0.77, 0.77]
B4	%	69	[0.69, 0.69]	B5	TLI	20	[0.8, 0.8]
B6	API	30	[0.94, 0.94]	B7	dB	73	[0.39, 0.39]
B8	qualitative	L3	[0.4, 0.6]	B9	qualitative	L2 secondary	[0.6, 0.8]
B10	Celsius degree	8	[0.8, 0.8]	B11	%	20	[0.67, 0.67]
C1	%	70	[0.7, 0.7]	C2	%	70	[0.7, 0.7]

(*continued*)

Table 4. (*continued*)

indicators	dimension	Index	Vague	indicators	dimension	Index	Vague
C3	Km/h	10	[0.75, 0.75]	C4	qualitative	L1	[0.8, 1]
C5	qualitative	L1	[0.8, 1]	C6	qualitative	L3	[0.4, 0.6]
C7	qualitative	L1	[0.8, 1]	C8	qualitative	L2	[0.6, 0.8]
D1	qualitative	L5	[0, 0.2]	D2	qualitative	L1	[0.8, 1]
D3	/	0.5	[0.83, 0.83]	D4	Quantity /km	0.2	[0, 0]
D5	qualitative	L1	[0.8, 1]	D6	%	1.25	[0.18, 0.18]
E1	%	72	[0.72, 0.72]	E2	qualitative	L1	[0.8, 1]
E3	qualitative	L3	[0.4, 0.6]	E4	qualitative	L2	[0.6, 0.8]

4.2 Vague Set of Objects to Be Evaluated

Vague value of vague value and corresponding value of ecological road evaluation system established in this study are shown in Table 4 for the North highway (construction highway ~ Planning Panlong Highway) section:

4.3 Calculate the Similarity Between Each Index and Each Level of the Object to Be Evaluated

The similarity is calculated according to formula (10), where v_i represents the index value of the item i to be evaluated, and w_{ij} represents the index value corresponding to the evaluation level j, $(j = 1, 2, 3, 4, 5)$ of the item i.Through calculation, the similarity between each index of the object to be evaluated and its five evaluation levels can be obtained. The greater the similarity is, the closer it is to the corresponding evaluation. The settlement result is shown in Table 5 below.

Table 5. Calculation results of the similarity between each index of the evaluation object and each evaluation level

	$Z(v_i, w_{i1})$	$Z(v_i, w_{i2})$	$Z(v_i, w_{i3})$	$Z(v_i, w_{i4})$	$Z(v_i, w_{i5})$
A1	0.034	0.034	0.031	0.028	0.017
A2	0.027	0.029	0.028	0.026	0.019

(*continued*)

Table 5. (*continued*)

	$Z(v_i, w_{i1})$	$Z(v_i, w_{i2})$	$Z(v_i, w_{i3})$	$Z(v_i, w_{i4})$	$Z(v_i, w_{i5})$
A3	0.021	0.022	0.023	0.022	0.012
A4	0.020	0.024	0.022	0.021	0.014
A5	0.026	0.023	0.020	0.017	0.014
A6	0.021	0.024	0.022	0.021	0.016
A7	0.025	0.029	0.025	0.022	0.018
B1	0.025	0.027	0.025	0.024	0.011
B2	0.030	0.031	0.032	0.031	0.020
B3	0.033	0.034	0.034	0.028	0.021
B4	0.031	0.037	0.036	0.029	0.025
B5	0.032	0.034	0.031	0.029	0.021
B6	0.037	0.037	0.033	0.028	0.017
B7	0.014	0.021	0.022	0.023	0.019
B8	0.019	0.022	0.025	0.022	0.019
B9	0.022	0.026	0.022	0.019	0.016
B10	0.020	0.022	0.019	0.017	0.015
B11	0.028	0.031	0.033	0.030	0.022
C1	0.016	0.020	0.018	0.016	0.015
C2	0.016	0.020	0.018	0.016	0.015
C3	0.018	0.021	0.020	0.018	0.012
C4	0.024	0.021	0.019	0.016	0.013
C5	0.022	0.020	0.017	0.014	0.012
C6	0.018	0.021	0.024	0.021	0.018
C7	0.022	0.020	0.017	0.014	0.012
C8	0.024	0.027	0.024	0.021	0.017
D1	0.015	0.019	0.022	0.026	0.030
D2	0.024	0.021	0.018	0.015	0.012
D3	0.022	0.022	0.023	0.018	0.012
D4	0.012	0.013	0.011	0.014	0.016
D5	0.029	0.025	0.022	0.018	0.015
D6	0.012	0.016	0.021	0.024	0.023
E1	0.022	0.024	0.025	0.026	0.014

(*continued*)

Table 5. *(continued)*

	$Z(v_i, w_{i1})$	$Z(v_i, w_{i2})$	$Z(v_i, w_{i3})$	$Z(v_i, w_{i4})$	$Z(v_i, w_{i5})$
E2	0.024	0.021	0.018	0.015	0.012
E3	0.017	0.019	0.022	0.019	0.017
E4	0.022	0.026	0.022	0.019	0.016

According to the above calculation results, Formula (11) is used to calculate the similarity levels of each category for different categories, and the calculation results are shown in Table 6.

Table 6. Results of classification evaluation

classification	1	2	3	4	5	level
Conservation of resources	0.005062	0.005373	0.005027	0.004541	0.0032	L2
Ecological protection	0.009851	0.010784	0.010526	0.009371	0.006878	L2
Construction of quality	0.003816	0.004041	0.003705	0.003229	0.002699	L2
Smart Management	0.002995	0.003077	0.003105	0.003046	0.002903	L3
Human Services	0.002089	0.002198	0.00215	0.001958	0.001453	L2

It can be found that the intelligent management of this section is relatively inadequate as Level 3, while the development of other fields is relatively good as Level 2. The main reason is that this section, considering that the design of intelligent transportation system is a strong overall and systematic work, pretends to cancel the setting of intelligent transportation facilities in the self-construction of some sections on Chongming Island, waiting for the overall planning scheme of subsequent departments.

4.4 Comprehensive Evaluation of Ecological Roads

According to formula (11), the similarity Z_j between the object to be evaluated and the evaluation level j was calculated.w_i is the weight of the index i. Specific calculation results are shown in Table 7.

Table 7. Calculation results of similarity between evaluation objects and each evaluation level

Z_1	Z_2	Z_3	Z_4	Z_5
0.823961	0.881359	0.847127	0.766883	0.599489

Based on Vague - Extenics ecological road evaluation method, similarity between evaluation sections and each evaluation level was calculated: $Z_2 > Z_3 > Z_1 > Z_4 > Z_5$.

So the comprehensive evaluation level of ecological road along highway (construction highway ~ planning Panlong highway) in the north was "Level 2", which was consistent with the fact.

5 Conclusion

This study analyzed the existing ecological road evaluation methods, compared the advantages and disadvantages of various evaluation methods, and analyzed the needs of ecological road evaluation. The evaluation index system of ecological road was obtained and established by consulting engineering design books and expert scoring method, and the weight of ecological road evaluation index was determined based on free structure entropy weight method. Based on Vague set theory, an ecological road evaluation model based on Vague- extension was proposed. Finally, based on the Vague vague of assessment index standard and assessment object, the comprehensive assessment level of ecological road in Chongming District of Shanghai (construction highway ~ Planning Panlong Highway) is determined as the second level by combining the vague of assessment index standard and assessment object. It was proved that the ecological road evaluation method based on Vague - extenics in this paper could still deal with different kinds of indicators comprehensively and complete the qualitative and quantitative comprehensive evaluation when the data is relatively simple.

Acknowledgments. The project is supported by Science and Technology Commission of Shanghai Municipality (Number 22dz1208505, 20DZ1202900, 19DZ1204200).

References

1. Wu, Q.: "American Recovery and Reinvestment Act of 2009" and its implications for our nation. Jiangxi Soc. Sci. **7**, 186–190 (2010)
2. The French National Assembly. Traffic Future Guidance Act (2019)
3. Ministry of Transport of the People's Republic of China. The Implementation plan for promoting ecological progress in transportation (2017)
4. Kuang, X.: Application of green ecological road concept in road construction of sponge city. China Construction **21**, 138–139 (2019)
5. Qi, W., Zhao, J.: Study on evaluation system of green ecological road. Hunan Commun. Sci. Technol. **47**(01), 12–15+26 (2021)
6. Chen, Y.: Data-oriented Safety Evaluation Method on CBTC Zone Control System. Tongji University (2019)
7. Xiong, Q.: Research on Key technologies of Multi-dimensional evaluation of new rail transit equipment. Tongji University (2021)
8. Fang, Y.: The Research on Safety Mechanism and Assurance Theory of Train Control System Based on 5M Model. Tongji University (2019)
9. Liu, H.: Study on Optimization of Fresh and Cold Chain Logistics System of Shandong Aviation Logistics Co., Ltd. Under the Background of New Retail. Civil Aviation University of China (2020)
10. Zang, Z., Cui, C.: Research on knowledge-based recommendation by one-dimensional of properties and requirement matching based on vague sets. Oper. Res. Manage. Sci. **29**(08), 112–119 (2020)

11. Wang, W., Wu, Q.: Analysis on the score function in vague set theory. Trans. Beijing Inst. Technol. (04), 372–376 (2008)
12. Zhao, J., Li, F.: Similarity measurement of vague set (Value). Comput. Eng. Appl. **55**(15), 69–74+103 (2019)
13. Chen, C., Lv, Z., Qin, P.: New method to measure similarity between intuitionistic fuzzy sets based on normal distribution functions. J. Chin. Comput. Syst. **28**(03), 500–503 (2007)

Extension and Calibration of BPR Function for Traffic Efficiency Based on MFD

Wei Shen[1], Xiang Liu[1,2](✉), and Fasheng Liu[2]

[1] School of Automotive Studies, Tongji University, Shanghai 200092, China
liuxiang@tongji.edu.cn

[2] College of Information and Control Engineering, Shandong Vocational University of Foreign Affairs, Weihai, China

Abstract. Considering temporal and spatial characteristics of traffic conditions, traffic efficiency is defined to indicate capacity of urban network. First the BPR volume delay function is expanded as a travelling speed function of traffic density instead of travelling time function of traffic volume, and then the capacity for a network is defined. The parameters are calibrated connecting with traffic parameters such as free speed, optimal speed and capacity. The macroscopic fundamental diagram (MFD) can deal with the relations among the traffic speed and densities of the road network in some scales. While it doesn't relate the density spread situation to traffic efficiency well and a generalized macroscopic fundamental diagram (GMFD) relating to density spread is proposed. It is found that average speed is a nonlinear function of average density and density spread at urban network level. Considering the density spread, traveled speed function is constructed to indicate traffic efficiency of network for travelers, based on the extended BPR (Bureau of Public Road) function from a road segment to the whole urban network. An improved Levenberg-Marquardt (LM) algorithm is used to calibrate parameters of the function and an improved Levenberg-Marquardt (LM) algorithm is used to calibrate parameters of the function. From case of Taidong business district, it proves that traveled speed function of network considering density spread is more in line with the physical variation of actual traffic and the calibration method is reliable.

Keywords: macroscopic fundamental diagram · traveled speed function · traffic efficiency · network capacity Levenberg-Marquardt algorithm

1 Introduction

Traffic congestion is a typical urban disease bothering traffic managers. Limited space resources of the city and existence of Brass paradox indicate that merely increasing in construction of transportation infrastructure does not solve the traffic problem, therefore, traffic control is an important measure for traffic management. However, traffic control always relies on forecasting models of traffic demand. From "four-stage" model in the 1950s, network equilibrium model and the disaggregate model in the 1970s to dynamic simulation model in the 1990s, they all hold certain defects [1, 2]:

X. Zeng et al. (Eds.): ITASC 2022, LNEE 1042, pp. 39–54, 2023.
https://doi.org/10.1007/978-981-99-2252-9_4

- calibration of the model requires many parameters, such as acquisition of dynamic od matrix;
- path selections of travelers are diverse and subjective, which is a complex game, unpredictable;
- behaviors of the oversaturated network are in chaos.

The emergence of macroscopic fundamental diagram (MFD) solves this problem well. The theory describes the relationship between average flow, average density and average speed of urban network, the stable and reproducible function relationship between traffic variables at network level, also reflects the proportional relationship between inflow and out flow of the network. The data type of MFD is single with not complicated data processing, therefore, it is suitable for large-scale and short-time monitoring state of road network, and widely applied in macro control of traffic.

Traffic breakdown is a common phenomenon in the real traffic system when traffic demand reaches or approaches the traditional traffic capacity where traffic flow is no longer continuous. Capacity of traffic network is not a simple addition of link flow, and it is difficult to be calibrated based on local flow observations. The era of big data brings new opportunities to the revolution of traffic theories and calibration of network capacity. One of the characteristics of big data is realizing parallel transmission of real-time data. Through recordings of moving vehicles and GPS information, vehicle location information of temporal and spatial are easily obtained. The full-scale macro data and methods of data processing under big-data condition can offset local micro-randomness to some extent. It is a new way to combine capacity (maximum traffic efficiency) extracted from big data with traffic management and control for urban traffic. For travelers on the urban network, traffic efficiency can be expressed by volume delay function, which is easily to be perceived.

This paper is organized as follows. Section 2 reviews the mostly related work on the topic of this paper, such as the extension of BPR and properties of GMF. Section 3 establishes traveled speed function of urban network for interpreting traffic efficiency for travelers, and proposes an improved Levenberg-Marquardt (LM) algorithm to calibrate parameters of the function. Section 4 defines research area and explores the effect of spatial distribution of density with real data from Taidong business district of Qingdao (China), and proposes a generalized macroscopic fundamental diagram (GMFD) based on traffic efficiency. Dataset from Taidong business district of Qingdao (China) is used for model test and error analysis on calibration results. Section 5 provides summary and discussion for further study.

2 Related Work on MFD and Nonlinear Function Fitting Review

2.1 Research Work on BPR and MFD

Volume delay function (VDF) reflects road characteristics, which affects volume distribution, relationship between user equilibrium and system optimization. The well-known BPR volume delay function is originally defined on a road.

$$T = L[a + (\alpha Q)^{\beta}] \quad (1)$$

T is the traveled time, L is the total length of the link on the road network, and Q is the traffic volume on the link. While a, α, β are parameters determined both by observations and analysis.

Through introduction of complex driving concepts, the relationship between parameters of volume delay function and road characteristics (such as free speed, optimal speed and optimal density) is theoretically deduced.

Take density spread into consideration, the general form of BPR function is as following:

The relations between the parameters and the traffic parameters of characteristics of the road i.e. free speed, optimal speed and optimal density on the road have been got by analysis [1].

$$\alpha=\frac{1}{v_f},\ \beta = \frac{1}{2v_m} \tag{2}$$

and

$$V_f = \frac{1}{\alpha},\ V_m = \frac{1}{2\beta} \tag{3}$$

Volume delay function is only defined on a link and it can't deal with the traffic situation for a road network. The traffic volume is essentially a local concept on a link. It can have an extension form of function of density instead of volume for road network.

The relationship of speed-density and flow-density has been the basis of traffic flow theory. The concept on macroscopic fundamental diagram (MFD) of traffic network is first proposed by Daganzo [2]. It mainly describes the relationship between average flow, average density and average speed from a network perspective. Especially the diagram of average flow and density. In the later years, some researchers aimed to explain the existence of MFDs with real data and simulations [3–5].

The factors that influence the attributes of MFD has been also examined. Studies of Geroliminis [4] show that the influencing factors of MFD mainly include four aspects: road facilities, traffic conditions, control conditions and behaviors of path selection. Essentially, all the factors influence the change in density of network. The spatial distribution of density is a key factor to obtain a well-defined MFD ignoring strict homogeneity of traffic states [6–8]. In process of exploring inhomogeneity of density, some researchers found the hysteresis phenomena in MFD [9, 10]. Then a three-dimensional macroscopic fundamental diagram (GMFD) for urban networks considering density spread was proposed [11, 12], and the relationship between macro traffic variables is redefined by some literatures considering stochastic characteristics and equilibrium of urban network [13–16]. Mazloumian et al. proved that the aggregated flow is a function of aggregated density and its spatial variation [7]. Some other researchers explored the influence of mechanisms of traffic signal (such as adaptive traffic signal systems) [17, 18] and the influence of driver routing on the shape and reliability of MFDs [19, 20].

Traffic benefits has temporal and spatial characteristics, and the distribution of density is a key factor in influencing the relationship between speed-density on a network. Macro traffic management is built based on accurate real-time information. To deal with the heterogeneous traffic distributions, networks are usually partitioned into homogeneous regions to get more reliable MFDs considering the variance of link densities and spatial compactness of partitioned regions [21]. Due to easily available trait and function of monitoring traffic states in real time, MFD are widely used in traffic control [22, 23] and cordon pricing [24] The era of big data brings new way to acquire MFDs of cities. Data mining approaches are applied for creating fundamental diagram of traffic flow [25] which makes MFD more easily available.

In summary, the afore mentioned studies suggested spatial spread of density influences the shape of MFD and MFD created by big data hold temporal and spatial characteristics of traffic conditions. However, studies on relationship between macro traffic variables neglect the temporal and spatial characteristics and parameters of fitted function has no actual meaning of urban network. As we later discuss, the effect of spatial distribution of congestion is still investigated based on traffic efficiency, and volume delay function is used for fitted function.

2.2 Traffic Capacity of Road Network

In this paper, traffic efficiency is defined to describe the traffic performance. Traffic network capacity is put as the maximum number of vehicle kilometers which a network can handle, where vehicles are evenly distributed with optimal speed. The formulation is calculated as following:

$$G_E = \sum_{i=1}^{n} N_i V_{si},\ C = \max\{G_E(T)\} \tag{4}$$

G_E stands for the traffic efficiency of a traffic network, C is the capacity of a traffic network, which can be directly extracted by long term statistic of traffic data, T is the time dimension, N_i stands for the total number of vehicles on link i in units of pcu, V_{si} is the space speed of link i in units of km/h, n stands for the total number of links on the traffic network.

If the average density $\overline{\rho}$ and total length L of links are given, the traffic efficiency can be calculated as below:

$$G_E = \overline{\rho} L \overline{V} \tag{5}$$

The average density (24), average weighted flow and average speed of the network comprised n links is determined as the following:

$$\begin{gathered} \overline{\rho} = \frac{\sum_{i=1}^{n} (\rho_i l_i n_i)}{\sum_{i=1}^{n} (l_i n_i)},\ \overline{Q} = \sum_{i=1}^{n} (q_i l_i n_i) / \sum_{i=1}^{n} (l_i n_i) \\ \overline{V} = \sum_{i=1}^{n} (Q_i l_i n_i) / \sum_{i=1}^{n} (\rho_i l_i n_i) \end{gathered} \tag{6}$$

ρ_i stands for the density of link i, q_i stands for the flow of link i, l_i stands for the length of link i, n_i stands for the lanes of link i.

Considering the characteristics of inhomogeneous distribution of network density, the traffic efficiency varies with the density variance of different links. The spatial spread of density is estimated by the square root of the weighted variance of densities in all links as shown below (7)

$$\gamma = \sqrt{\frac{\sum_i l_i(\rho_i - \overline{\rho})^2}{\sum_i l_i}} \tag{7}$$

3 Mathematical Analysis for the Extensions

In this section, traveled speed function is established to describe traffic efficiency for travelers. Considering the spread of density, BPR (Bureau of Public Road) function of a link is extended to the whole network. Levenberg-Marquardt (LM) algorithm is proposed to calibrate parameters of the function, and data of Taidong business district prove this model reliable.

3.1 Establishment of Traveled Speed Function

Traffic efficiency of road network contains spatial and temporal properties. Traveled speed function can provide a quantitative description and perception of traffic efficiency for travelers.

According to Eq. (5), traveled speed function $V = V(\rho, \gamma)$ related to density is an implicit function relation:

$$\overline{V}^{-1} = a + [b(\overline{\rho} + \gamma)\overline{V}]^c \tag{8}$$

Combining with basic theory of traffic flow:

$$\begin{cases} \overline{V}(0, 0) = V_f \\ \overline{V}(\rho_m, \gamma_m) = V_m \\ \overline{V}''(\rho_m, \gamma_m) = 0 \\ \gamma_m = 0 \end{cases} \tag{9}$$

We can deduce the relationship between flow delay function parameters and road characteristics $a = 1/V_f$ and $b, c \sim (V_f, V_m, \rho_m)$. Combined with Eq. (4) and (5) obtain the optimal traffic efficiency of road network, which is determined by network characteristics:

$$G_{Em} \sim (a, b, c, L) \tag{10}$$

Through the traffic big-data platform, we can directly obtain real-time data on traffic efficiency, and peak value of certain time window T is corresponding to network capacity (maximum traffic efficiency), where the vehicle speed and traffic density is also at the optimal level. By Eq. (6), we can establish evaluation system of traffic efficiency among different cities. The relationship between parameters of traveled speed function and network characteristics can be used to calibrate relevant parameters.

3.2 Calibration Method

Calibration of parameters of traveled speed function is transformed into a problem of nonlinear function fitting. Levenberg-Marquardt (LM) algorithm is a commonly method.

However, the traditional LM algorithm fails to solve the nonlinear least squares under large residual. In this paper, a self-optimizing LM algorithm that the residual changing direction is used to search for trust-region intervals is proposed. The updating algorithm of damping coefficient is improved. A judging condition is introduced to solve the local divergence caused by large residual, and the LM algorithm is substituted by the steepest descent method. It is the unique accepting condition that the objective function value is decreased continuously in the iterative process. So the self-optimizing LM algorithm is stable and reliable.

The least squares objective function $F(x, w)$ based on the square sum of estimation (SSE), expressed as:

$$F(x, w) = \sum_{j=1}^{M} \sum_{i=1}^{N} f_{i,j}^2 \tag{11}$$

x is the input parameter vector,w is unknown parameter variables, which needs calibration. N stands for the number of unknown parameter variables, f is the single residual value of objective function, M stands for the number of sampling points in discrete nonlinear system.

The iteration formula of LM algorithm is as shown below:

$$w_{k+1} = w_k - (J_k^T J_k + \mu I)^{-1} J_k e_k \tag{12}$$

k is the iteration step, w_{k+1}, w_k stand for values of parameter searched at iteration step $k+1$ and k respectively, when the request parameter is w_k, J_k stands for Jacobian matrix, f_k stands for the single residual value, e_k represents the unit vector group, and I represents the unit matrix.

Solution Algorithm is as follows:

1. Initialize the variable $k = 0$, define the initial value w_0, and stop the iteration for condition ε and maximum step $k_{\max}$.
2. Calculate Hessian matrix $H(H = J_k^T J_k)$ and gradient vector g, $(g = J_k f_k)$.
3. Initially, steepest line of descent is used to calculate descending direction h, and one-dimensional linear search to calculate the exact step size d to get the search point w_1, $w_1 = w_0 - dh$. Add iteration step, $k = k + 1$, then update search point $w_0 = w_1$ and calculate H and g.
4. If $\|h\|_2 < \varepsilon$, go to step 11. Otherwise, calculate the target function F_0, set $\lambda_0 = 0$, $\lambda_a = 10$, scale factor $v = 2$, iteration interruption symbol $\rho = 0$, residual increase symbol $l = 0$, search count variable $num = 1$.
5. Calculate variation h of expected parameter, $h = -H^{-1}g$, then update search point $w_1 = w_0 - h$, calculate objective function F, and set $\lambda_{\min} = \lambda_0, \lambda_0 = \lambda_0 + 2\lambda_a, \lambda_{\max} = \lambda_0$.
6. If $F_1 \le F_0$, update search point $w_0 = w_1$ and iteration steps $k = k+1$. If $k > k_{\max}$ then go to 11. Otherwise, go to 4; if $F_1 > F_0$, set $F_{\min} = F$.

7. Update damping coefficient $\mu = (\lambda_{min} + \lambda_{max})/2$, $F_{old} = F_1$ and matrix H,$H = J_k^T J_k + \mu \times diag(J_k^T J_k)$, then calculate variation h of expected parameter, update search point $w\prime = w_0 - h$, and calculate objective function F_1.
8. If $F_{min} > F_1$, set $F_{min} = F_1$, $\lambda_{min} = \lambda_0$, $num = 0$. Otherwise,$num = num + 1$, if $num > 10$, set $\rho = 1$, go to 9. Otherwise, go to 10.
9. Steepest descent method is used to replace this iterative process. Calculate iterative search point w_1, $w_1 = w_0 - dh$, update search point $w_0 = w_1$, go to 4.
10. If $F_{old} \leq F_1$, the search direction residuals will be increased, set $\lambda_{max} = \lambda_0$, $\lambda_{min} = \lambda_0$, $\lambda_0 = \lambda_0 + \lambda_a/5$, $l = 0$, $v = 2$. If $F_{old} > F_1$, the search direction residuals will be reduced, if $l = 1$, set $v = 2v$. Update variables $\lambda_{min} = \lambda_0$,$\lambda_0 = \lambda_0 + \lambda_a \times v$, $l = 1$, $\lambda_{max} = \lambda_0$. Jump to 6.
11. End of iteration.

4 Data Derivation and Fitting for Taidong District of Qingdao

4.1 Definition of Research Area

This section provides the information of studied road network and data derivation of traffic properties. Taidong business district of Qingdao is chosen to describe characteristics of MFD. The structure of road network including road grade, number of lanes and road length influences the shape of MFD, so the detail of Taidong business district and its data collection is shown in this section.

Taidong business district is a famous shopping district in Chinese. The traffic congestion phenomenon often occurs in the morning and evening peaks there, so it is easier to obtain a more complete MFD of traffic flow. The location of Taidong business district is shown in Fig. 1. This place consists of an area of 60.2872 hectares and a high-resolution navigation network including 11 road segments (4 arterial roads, 4 collector roads and 3 branches). Total length of studied links is 9.72 km. Table 1 provides specific information of all the road segments. Due to the road network consists of different types of road, it hardly satisfies the homogeneity conditions proposed by Geroliminis and Daganzo [4] to obtain a well-defined network fundamental diagram. However, we emphasized on the property of density spread into analysis instead of dividing optimal partitioning of network regions characterized by homogeneous traffic conditions.

Table 1 shows the length of road segments, the number of lanes and road grade in Taidong business district. Yan'an Road and Weihai Road are the two arterial roads, and the number of two-way lanes of Weihai Road is different, so their statistical data are divided into two parts according to the direction of the road section. This method makes spatial characteristics of traffic flow more accurately grasped. The information of Taidong business district implies that there will be more scatters in MFD, because the shape of road network is irregular and its road grade is not uniform, which result in uneven density distribution.

In Qingdao city of China, fixed vehicle detectors covers artery roads, collector streets and key branches. The information of average speed, time occupancy, passed vehicles and other key attributes of traffic state are detected and uploaded every minute to the data center. Then the traffic network properties (traffic efficiency, weighted average density and average speed etc.) can be calculated based on these data. Data can be obtained from the traffic information service website of public security bureau of Qingdao, which is available in China (Qingdao, which is available in China (Source: http://www.qdznjt.com /trafficIndex/real Time Index#). Dataset includes traffic volume, density and speed of vehicles). Dataset includes traffic volume, density and speed of vehicles in 11 roads of Taidong business district and covers a whole day (24 h). The statistic is aggregated every five minutes, which can monitor traffic conditions of the whole network in short periods.

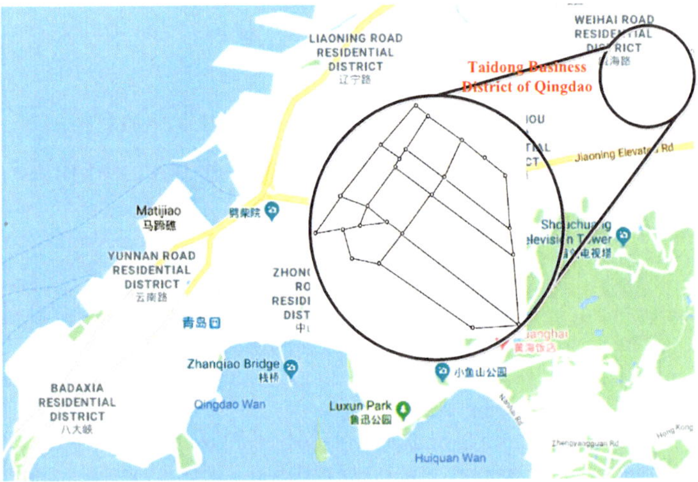

Fig. 1. Taidong business district of Qingdao (adapted from Google Map).

Table 1. Road information of Taidong business district

Number	Segment name	Road grade	Length of section (m)	Lanes
1	Yan'an Road (South to North)	arterial road	1200	2
2	Yan'an Road (North to South)	arterial road	1200	2
3	Sangzi Road	branch	702.6	2
4	Weihai Road (South to North)	arterial road	921	2
5	Weihai Road (North to South)	arterial road	921	3
6	Changchun Road	arterial road	682.8	6
7	Lijin Road	collector road	1000	4
8	Liaoning Road	collector road	459	4
9	The fist Taidong Road	arterial road	902	4
10	The sixth Taidong Road	collector road	914	2

(*continued*)

Table 1. *(continued)*

Number	Segment name	Road grade	Length of section (m)	Lanes
11	The eighth Taidong Road	collector road	825	2
12	Daokou Road	branch	363.5	1
13	Shunxing Road	branch	809	2

4.2 Effect of Spatial Inhomogeneity of Density

In this section, we investigate MFD properties of Taidong business district. The main peculiarity of this studied district consists of: (a) vast and highly heterogeneous urban network (including arterial roads, collector roads, and branches), (b) the spatial heterogeneity of origins and destinations, and (c) the realistic travel behavior of the data series which covers 24 h of one day. As previous literatures detected (9–10), high volume of scatters appear in the diagram of relationship between average weighted density and average weighted flow when the regularity conditions are relaxed. Besides, a correspondence between clockwise patterns is identified in the diagram.

4.2.1 Two Dimensional MFD

MFDs of Taidong business district are demonstrated in this section, and macro traffic variables in diagrams are calculated by formulation in Sect. 3.1. Traffic data is a time series aggravated every 5 min. Due to different daily traffic conditions, the shapes of MFD appear inconsistent. Besides, traffic efficiency holds temporal and spatial characteristics, which leads to diverse shapes of MFD under different time window and spatial inhomogeneity.

From Eq. (1), we can know that data in larger time span makes network capacity more accurate, so does the model of MFD. Figure 2 (a) shows the MFDs of Taidong business district based on a total of 1445 data in five days. Table 2 is the quantitative relationship between average density and traffic efficiency, which fitted by quadratic parabola. In Fig. 2(a), traffic efficiency on August 16, 2017 is poor. Even in the low-density state, the traffic efficiency appears many scatters, and the optimal value is also lower than other days. This phenomenon demonstrates that there is a serious traffic congestion on that day. Similarly, MFD for May 16, 2017 also show a lot of scatters, which lead to a smaller value of network capacity on that day, mainly due to uneven distribution of vehicles in road sections. However, MFD for April 16, 2017 containing plenty of scatters reach the maximum traffic efficiency (network capacity). On the contrary, MFD for March 21, 2017 hold a good shape, and traffic efficiency on this day reaches an optimal level. The maximum density appears on February 23, 2017.

Combining different shapes of MFD with fitted results, we can know that optimal density of Taidong business district is 55 pcu/km/*lane* fluctuating within 10 pcu/km/*lane*. This parameter is reference for macro management of road network. Correspondingly, the optimal traffic efficiency is $1300\,\text{pcu} \cdot \text{km}/lane$ fluctuating within $50\,\text{pcu} \cdot \text{km}/lane$. According to Eq. (4), we can know the network capacity of this district is $1423\,\text{pcu} \cdot \text{km}/\text{h}$ fluctuating within $50\,\text{pcu} \cdot \text{km}/lane$.

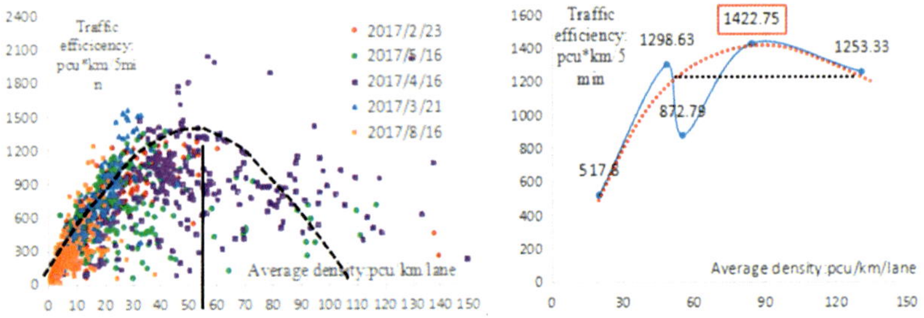

(a) MFD of density and traffic efficiency (b) curve of optimum value of fitted quadratic parabola

Fig. 2. MFD for Taidong business district

Table 2. Fitting result of macro relationship

Quadratic function of density- traffic efficiency relationship:

$G^E = a\overline{\rho}^2 + b\overline{\rho} + c$

Date	a	b	C	R-squared of density fit	Optimal average density	Optimal traffic efficiency
2017/2/23	−0.0553	14.491	304.01	0.2351	131.02	1253.33
2017/5/16	−0.1908	21.21	283.35	0.2486	55.58	872.79
2017/4/16	−0.1789	30.275	141.9	0.7553	84.61	1422.75
2017/3/21	−0.5503	53.762	−14.455	0.8008	48.85	1298.63
2017/8/16	−1.2865	51.797	−3.5665	0.4476	20.13	517.80

4.2.2 Generalized Macroscopic Fundamental Diagram

MFD for February 23, 2017 is used to make a further analysis on reason for scatters. The maximum traffic efficiency always appear at travel peak, and accompanied by typical phenomenon that time series of traffic variables(density, density spread and traffic efficiency) changed not synchronously. During this period, the hysteresis loop is easily found in MFDs.

Figure 3 (a) reveals the parabolic trend on relationship of traffic efficiency and average density. Figure 3 (b) shows the change pattern of macro variables with time, we found that average density and traffic efficiency vary consistently in most of the times when the road is in good traffic state. But there is still some converse change during peak hours of traffic when traffic efficiency approaches closer to network capacity, which can be seen in black circle of Fig. 3 (b). Enlarging black circle in Fig. 3 (b) to Fig. 3 (c) and Fig. 3 (d), it is very evident that the traffic congestion is not evenly distributed in spatial and temporal aspects. Some parts of urban network are more congested than others because of inhomogeneity. During the peak hours, traffic conditions are significantly affected by local traffic congestion, and even at low average densities, the traffic efficiency of the entire network is reduced. The instability of different traffic states (such as congestion and non-congestion) is related to the hysteresis phenomenon in the traffic flow.

The hysteresis phenomenon occurs in 19:05–19:35, traffic efficiency has an anti-clockwise hysteresis loop changed with density, while density spread has a clockwise hysteresis loop changed with density (seen in red circles of Fig. 3 (c) and Fig. 3 (d)). The change of hysteresis loop in other periods has the same direction (seen in red lines and black lines in Fig. 3 (c) and Fig. 3 (d)). The converse hysteresis loop indicates that the increase in average density leads to the increase of density spread and decrease of traffic efficiency. In actual situation, vehicles enter the network but choose different path resulting its uneven distribution on road network. Through this, we conclude that the reduction in traffic efficiency is caused by the increase in density spread. If we study details of density distribution of road segment, we can see that the spatial spread of density increases from about 18:00 and gradually reaches a very high level lasting until 19:40. After that, the number of vehicles dropped to a normal level, the congestion gradually dissipated, and the traffic flow returned to a smooth state.

As congestion spread wildly on the whole network, the distribution of density is decreased. There is more flat and broad distribution of density considering the time

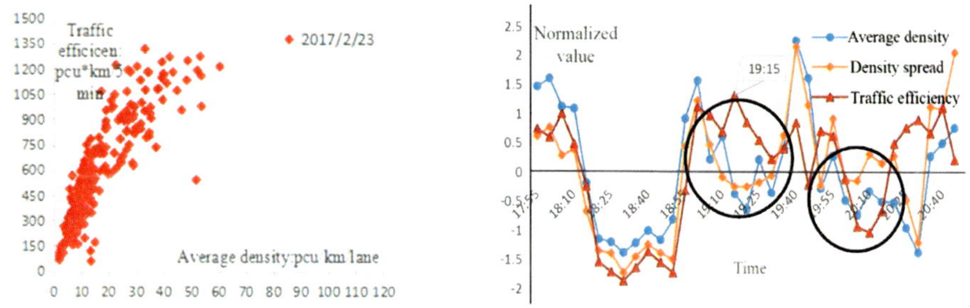

(a) MFD of Taidong business district on 2017/2/23, (b) Time-varying diagram of traffic variables during evening period

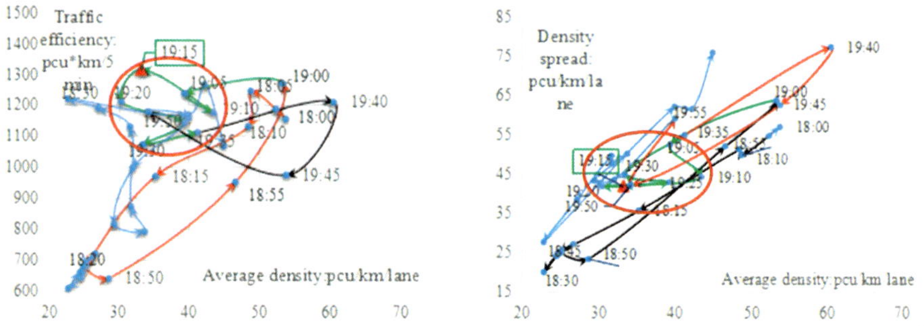

(c) Hysteresis loops in the density-traffic efficiency diagram, (d) Hysteresis loops in the density -density spread diagram

Fig. 3. MFD of Taidong business district on 2017/2/23

and space characteristics. Furthermore, the traffic efficiency and average speed both decline rapidly with increasing average density and density spread approaching a power function relation (seen in Fig. 3(a) and Fig. 3(b)). Since average speed of network can be converted into travel time, which is easier for the traveler to perceive, so it can be used as a measure of traffic efficiency of network. Significantly, for the same value of average density lower traffic efficiency and average speed can be both stimulated by a higher spread of density. This property is consistent with the hysteresis phenomena characterized by different levels of traffic efficiency at the onset and offset of congestion.

GMFD is extended from two dimensional MFD to three dimensional network fundamental diagram, which including traffic variables of network (density, density spread, average speed and traffic efficiency) shown in Fig. 4. A well-defined invariant MFD appears under circumstances of homogenous condition, however, this study shows high scattering in such circumstances (Fig. 2 (a)). The three-dimensional relation reveals that scatterings results from aggregated patterns instead of randomness, and they relates to spatial distribution of congestion.

Combining Fig. 4 (a) with Fig. 4 (b), we find that the inhomogeneous distribution of congestion influences the traffic efficiency even below critical densities, which can be seen in the range [30, 50] of average density. In some time intervals, drops in traffic efficiency occur together with rapid increase of density spread, which can be seen in the range [40, 50] of average density whereas the average density remains almost constant. We can easily deduce that, decrease in traffic efficiency generated by clusters of congestion rather than increase of travelers in the network. GMFD related to average speed of network appears many scatters at critical density (about 50 pcu/km/*lane*) illustrating traffic performance better. Due to inhomogeneity, travel conditions cannot be identified directly by MFD. However, it is important to understand the variation pattern followed before reaching the current point in the plane of density and traffic efficiency. So we need to consider spatial distribution of density to accurately describe traffic law in large-scale complex networks characterized by inhomogeneous traffic conditions.

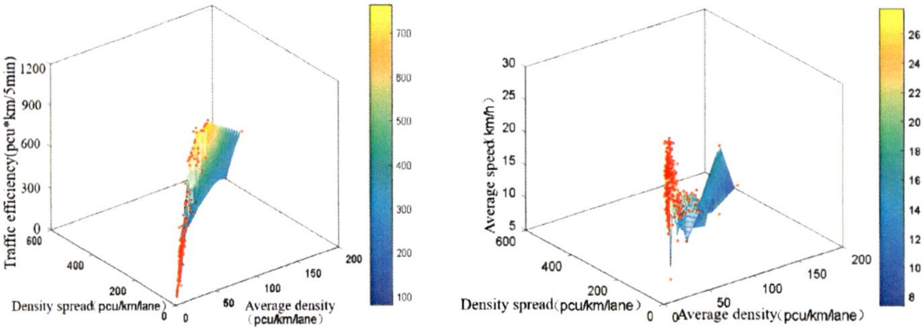

(a) GMFD of traffic efficiency, density and density spread, (b) GMFD of average speed, density and density spread

Fig. 4. GMFD of Taidong business district on 2017/2/23

4.2.3 Analysis of Calibration Results

Data series on 2017/2/23 of Taidong business district is used for calibration of parameters of traveled speed function. The front of dataset is used to calibrate the function, and the remaining of time series are used to test fitness of calibrated results. Through the error analysis, calibration method proves to be reliable.

The spatial spread of density can be calculated by Eq. (7). A curve fitting software named 1stOpt (version 1.5) is used to realized self-optimizing LM algorithm and calibrate the traveled speed function in Eq. (8) (called function (a)). To explain the influence of spatial spread of density on speed, we compare Eq. (8) with the equation as shown below (called function (b)):$\overline{V}^{-1} = a+[b\overline{\rho}\overline{V}]^c$. The calibration result of function (a) parameters are: $a = 0.038$ $b = 0.00000135$, and $c = 0.752$, while parameters of function (b) are: $a = 0.039$ $b = 0.0000064$, and $c = 0.857$. Correspondingly, traveled speed function of Taidong business district is:

$$\overline{V}^{-1} = 0.038 + [0.00000135(\overline{\rho} + \gamma)\overline{V}]^{0.752} \; or \; t = 9.72\left\{0.038 + [0.00000135(\overline{\rho} + \gamma)\overline{V}]^{0.75}\right\}$$

$$\overline{V}^{-1} = 0.039 + [0.0000064\overline{\rho}\overline{V}]^{0.857}$$

There are subtle differences between calibrated parameters of function (a) and function (b). Due to consideration on the spatial spread of density, parameter b, c of function (a) both are smaller than function (b), which characterize the ascending velocity of power function.

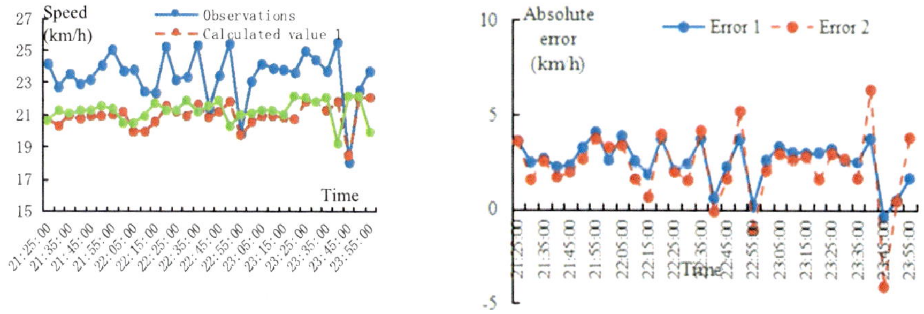

(a) Comparison between observations and fit value (b) Error analysis on fitted function (a) and fitted function (b)

Fig. 5. Analysis on suitability of fit results.

The error analysis of calibration results are demonstrated in Fig. 5. There is a comparison curve between calculated value and observations in sub-figure of 5(a). The curves of calculated value 1 and calculated value 2 are represents the calibrated values of function (a) and function (b) respectively. We can see that the fit curves almost close to observation curve, which means the calibration method is reliable. Calculated results of function (a) have the same trend as the observations, while function (b) with a closer value to observations. To explain this phenomenon, we need to combine this with the initial fit function, and function (a) takes the spatial spread of density into consideration, which plays a role in variation in average speed. Obviously, the spread of density makes traffic efficiency worse, which reflected in decreasing traveled speed of the whole network. However, in another aspect, it accurately describes the changing pattern of speed of network with increasing density. The speed of Taidong business district fluctuates fiercely in this statistical period, and the low speed reflects a terrible traffic condition at late night, and it maybe a special characteristic for business district which has a heavy trade during late night.

Absolute errors are shown in sub-figure of 5(b). Error 1 and Error 2 represents the calibrated error of fitted function (a) and fitted function (b) respectively. Both of their absolute errors fluctuate at value 2 except some extreme point which suffers worse traffic efficiency. When the absolute error is at a similar level, the function (a) is better than function (b) to describe temporal characteristics of traffic efficiency on a whole network, and its calibration errors fluctuate within acceptable range.

5 Conclusion

This paper defines traffic efficiency and capacity of urban network from perspective of total number of vehicle kilometers, and capacity of a network can be extracted directly from big statistic data, which can provide reference for traffic managers. MFDs of Taidong business district in Qingdao city was explored. It shows high volumes of scatterings appear in average density–traffic efficiency diagram. Besides, there is an anticlockwise hysteresis loop in density–traffic efficiency diagram while clockwise hysteresis loop in density–density spread diagram, which proves the fact that inhomogeneity of density leads to lower network capacity. From time series of traffic variables, it is found that the changing trend of traffic efficiency is shown as a single peak of density in good traffic conditions, and presents bimodal with a single peak of density during traffic congestion. The former appears more frequently than the later in most of the times, so only the density cannot describe the traffic state suitably for the inhomogeneity of density. Moreover, inhomogeneity should be discussed in considering temporal and spatial aspects.

A generalized macroscopic fundamental diagram (GMFD) relates average speed, density and density spread is proposed to illustrate traffic performance. Based on the nonlinear relationship of average speed to average density and density spread, volume delay function of road segment in BPR (Bureau of Public Roads) form is extended to the whole network, that is traveled speed function of network. Parameters of the function is calibrated by statistical method and theoretical derivation. Statistical calibration of parameters is through improved Levenberg- Marquardt algorithm. Data of Taidong business district in Qingdao is used for test. It is found that traveled speed function of network considering density spread is more in line with the physical variation of actual traffic and the improved algorithm is reliable. Given the optimal condition obtained from big data, the further study is effective measures to keep traffic state at optimal level.

References

1. Liu, X.: Analysis of EVC Traffic Benefits from a Perspective of the Unity Connection of User, Vehicle, Road and Flow; Ph D thesis Tongji University 2020
2. Daganzo, C.F.: Urban gridlock: macroscopic modeling and mitigation approaches. Transp. Res. Part B **41**(1), 49–62 (2007)
3. Daganzo, C.F., Geroliminis, N.: An analytical approximation for the macroscopic fundamental diagram of urban traffic. Transp. Res. Part B **42**(9), 771–781 (2008)
4. Geroliminis, N., Daganzo, C.F.: Existence of urban-scale macroscopic fundamental diagrams: some experimental findings. Transp. Res. Part B **42**(9), 759–770 (2008)
5. Hoogendoorn, S.P., Mario, C., Winnie, D.: Macroscopic fundamental diagram for pedestrian network. In: Transportation Research Board 89th Annual Meeting (2010)
6. Geroliminis, N., Sun, J.: Properties of a well-defined macroscopic fundamental diagram for urban traffic. Transp. Res. Part B **45**(3), 605–617 (2011)
7. Mazloumian, A., Geroliminis, N., Helbing, D.: The spatial variability of vehicle densities as determinant of urban network capacity. Philos. Trans. R. Soc. Math. Phys. Eng. Sci. **2010**(368), 4627–4647 (1928)
8. Buisson, C., Ladier, C.: Exploring the impact of the homogeneity of traffic measurements on the existence of macroscopic fundamental diagrams. Transp. Res. Rec. J. Transp. Res. Board **137**(2124), 127–136 (2009)
9. Geroliminis, N., Sun, J.: Hysteresis phenomena of a macroscopic fundamental diagram in freeway network. Transp. Res. Part A Policy Pract. **45**(9), 966–979 (2011)
10. Gayah, V.V., Daganzo, C.F.: Clockwise hysteresis loops in the macroscopic fundamental diagram: an effect of network instability. Transp. Res. Part B **45**, 643–655 (2011)
11. Geroliminis, N., Zheng, N., Ampountolas, K.: A three-dimensional macroscopic fundamental diagram for mixed bi-modal urban networks. Transp. Res. Part C Emerg. Technol. **42**, 168–181 (2014)
12. Knoop, V.L., Hoogendoorn, S.P.: Empirics of a generalized macroscopic fundamental diagram for urban freeways. Transp. Res. Board **2391**(1), 133–141 (2013)
13. Helbing, D.: Derivation of a fundamental diagram for urban traffic flow. Euro. Phys. J. B **70**(2), 229–241 (2009)
14. Wang, H., Ni, D., Chen, Q.Y., Li, J.: Stochastic modelling of the equilibrium speed–density relationship. J. Adv. Transp. **47**(1), 126–150 (2013)
15. Wang, H., Li, H., Chen, Q.Y., Ni, D.: Logistic modeling of the equilibrium speed–density relationship. Transp. Res. Part A **45**(6), 554–566 (2011)
16. Ramezani, M., Haddad, J., Geroliminis, N.: Dynamics of heterogeneity in urban networks: aggregated traffic modeling and hierarchical control. Transp. Res. Part B Methodological **74**, 1–19 (2015)
17. Zhang, L., Garoni, T.M., de Gier, J.: A comparative study of macroscopic fundamental diagrams of arterial road networks governed by adaptive traffic signal systems. Transp. Res. Part B **49**, 1–23 (2013)
18. Gayah, V.V., Gao, X., Nagle, A.S.: On the impacts of locally adaptive signal control on urban network stability and the macroscopic fundamental diagram. Transp. Res. Part B Methodological **70**, 255–268 (2014)
19. Mahmassani, H.S., Saberi, M., Zockaie, A.: Urban network gridlock: theory, characteristics, and dynamics. Transp. Res. Part C **36**, 480–497 (2013)
20. Saberi, M., Mahmassani, H.S., Zockaie, A.: Network capacity, traffic instability, and adaptive driving: findings from simulated urban network experiments. Euro. J. Transp. Logistics **3**(3–4), 289–308 (2013). https://doi.org/10.1007/s13676-013-0040-2

21. Ji, Y., Geroliminis, N.: On the spatial partitioning of urban transportation networks. Transp. Res. Part B **46**, 1639–1656 (2012)
22. Haddad, J., Ramezani, M., Geroliminis, N.: Cooperative traffic control of a mixed network with two urban regions and a freeway. Transp. Res. Part R **54**, 17–36 (2013)
23. Haddad, J., Mirkin, B.: Adaptive perimeter traffic control of urban road networks based on MFD model with time delays. Int. J. Robust Nonlinear Control **26**(6), 1267–1285 (2016)
24. Zheng, N., Waraich, R.A., Axhausen, K.W., Geroliminis, N.: A dynamic cordon pricing scheme combining the macroscopic fundamental diagram and an agent-based traffic model. Transp. Res. Part A Policy Pract. **46**(8), 1291–1303 (2012)
25. Kianfar, J., Edara, P.: A data mining approach to creating fundamental traffic flow diagram. Procedia Soc. Behav. Sci. **104**, 430–439 (2013)

Train Regenerative Braking Strategy Optimization Based on Reinforcement Learning

Xiaoqing Zeng[1], Liqun Liu[1], and Tengfei Yuan[2,3](✉)

[1] The Key Laboratory of Road and Traffic Engineering, School of Transportation Engineering, Ministry of Education, Tongji University, 4800 Cao'an Road, Shanghai, China
[2] SHU-UTS SILC Business School, Shanghai University, Shanghai 201800, China
yuantengfei@shu.edu.cn
[3] Shanghai Engineering Research Center of Urban Infrastructure Renewal, Shanghai 200032, China

Abstract. With the increasement of urban rail transit operation density, the power consumption of metro system is also rising sharply. Meanwhile the proportion for urban rail transit of power consumption is increasing, so this problem needs more and more attention. In order to reduce the power consumption of rail transit, this research mainly focuses on the renewable energy utilization of train, which means that the train will make the best of the regenerative braking energy. For this purpose, the flywheel energy storage device is used as on-board device, then the regenerative braking strategy of the train is optimized based on reinforcement learning algorithm. Ultimately, the optimized train speed curve by the dynamic planning and Q-learning can achieve more than 5% energy recovery of the total energy consumption. The results show that this research can save the power consumption of rail transit by recycling the braking energy, which is of great significance for significance for energy saving and green transportation

Keywords: Regeneration energy · Dynamic planning · Reinforcement learning · Strategy optimization

1 Introduction

Urban rail transit is a kind of transportation mode with high reliability and large transportation capacity. With the improvement of the economic development of cities at all levels and the continuous improvement of engineering technology, subway, as a typical representative of urban rail transit, has been growing rapidly in the whole country. At present, China has ranked the first in the world in terms of operating mileage and passenger flow scale. More and more attention has been paid to the research of energy-saving operation of rail transit.

The main research goal of train energy-saving driving is to find the optimal speed curve when the train energy consumption is the lowest through the analysis of the speed curve of the train on the section under the condition of meeting the constraints of the operation environment. The optimal speed curve is mainly determined by the train force

X. Zeng et al. (Eds.): ITASC 2022, LNEE 1042, pp. 55–66, 2023.
https://doi.org/10.1007/978-981-99-2252-9_5

and the tangent sequence of working conditions, so there are many research results at home and abroad. Kunihiko [1] solved the train speed curve as a discrete bounded state variable problem by using the Pontryagin maximum principle. Howlett [2] et al. Proved the existence of single train energy-saving optimal operation method. Oshima [3] further applies fuzzy control theory to train operation control and improves the punctuality rate and stopping accuracy of automatic train driving system (ATO). Masafumi [4] et al. Introduced the consideration of the state of charge (SOC) of the energy storage device when studying the optimal control model, and comprehensively solved the optimal SOC control strategy. DOMínguez [5] et al. Studied the energy-saving control driving strategy when the train is equipped with ATO, and proposed a model solution based on multi-objective particle swarm optimization algorithm. Bao [6] and others put forward the simplified principle of train state space based on Pang's maximum principle. Liu [7] et al. used the value function approximation method to estimate the optimal value function, which improved the accuracy and operation efficiency of train operation strategy. Wu [8] et al. Combined with the control strategy of on-board energy storage device, studied the optimization of train speed curve. They used mixed integer linear programming (MILP) to solve the model.

Regenerative braking is also called feedback braking or regenerative braking in the field of rail transit. The biggest feature is to use the reversibility of the motor. When the train is braked, the traction motor reversely acts as the output energy of the generator, which is usually called regenerative energy. Renewable energy has great practical role and research value in many fields such as electric vehicles. In rail transit, the use of traction power supply system can achieve multi vehicle cooperation, thus reducing energy consumption at the system level. The principle of regenerative braking is shown in Fig. 1:

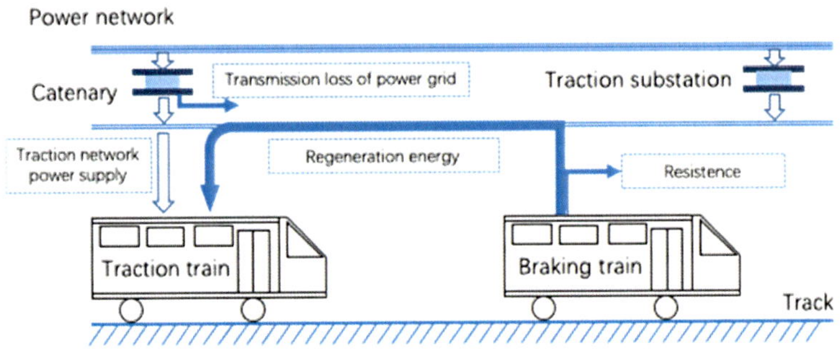

Fig. 1. Schematic of Direct feedback of train regenerative energy

The function of on-board energy storage device is to directly recover and store the regenerative energy generated by the train during braking, rather than feedback the traction network [9, 10]. Therefore, the on-board energy storage device can be used as an auxiliary power source to reduce the overall energy consumption of the traction power supply system under the condition of train traction. The schematic diagram of on-board energy storage device is shown in Fig. 2:

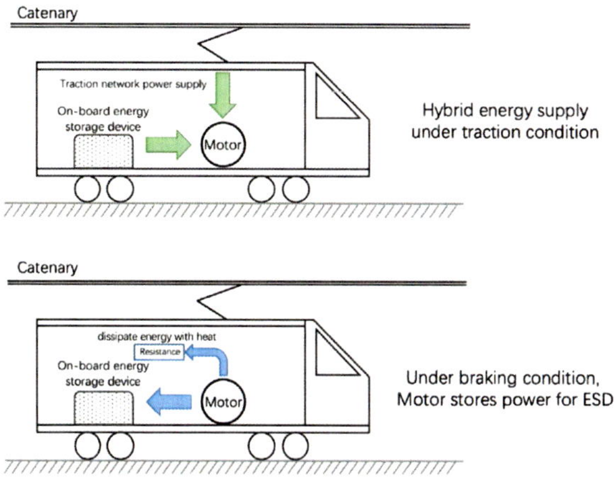

Fig. 2. Renewable energy utilization of train energy storage device

The purpose of this study is to analyze the optimization of the operation curve of the on-board energy storage train considering the recovery and utilization of regenerative braking energy, and to analyze and solve the evaluation of the optimal control curve of the train.

2 Model Formulation

When analyzing the forces on the running process of the railway train, it mainly needs to analyze the traction force, and running resistance of the train. It is very important for the calculation and optimization of the optimal speed curve to solve the stress state in different operation stages or working conditions. The train running resistance is generally divided into basic running resistance and additional running resistance, and the force situation is shown in Fig. 3:

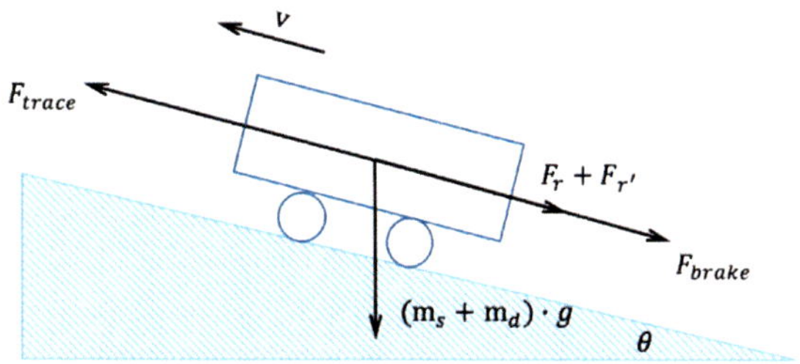

Fig. 3. Schematic of Single train force analysis

$$F_{trace}(v) = \begin{cases} F, & 0 \leq v \leq V_1 \\ \frac{P_0}{\mu_0 v + \varepsilon_0}, & V_1 \leq v \leq V_2 \\ \frac{P(v)}{\mu_1 v^2 + \mu_2 v + \varepsilon_1}, & V_2 \leq v \leq V_{max} \end{cases} \tag{1}$$

$$F_r = A + B \cdot v + C \cdot v^2 \tag{2}$$

$$F_{r\prime} = M \cdot g \cdot \sin\theta = M \cdot g \cdot i \cdot 1000 \tag{3}$$

where F is the constant traction output under the constant torque mode of the train, P_0 is the power of the train traction motor at constant power output, $P(v)$ is the output power of traction motor decreases with the increase of vehicle speed in the power reduction stage, μ_0, μ_0, μ_0 and ε_0, ε_1 are all represent fitting coefficient.

The basic resistance can be expressed by Davis Equation, and the main sources of additional resistance are: tunnel, turning radius and ramp. In this study, the ramp resistance is mainly considered. The resultant force on the train is indicated by $C(v)$:

$$C(v) = \begin{cases} \mu_f F_t(v) - \omega(v) - \omega\prime(x) \\ -(\omega(v) + \omega\prime(x)) \\ \mu_b F_b(v) - \omega(v) - \omega\prime(x) \end{cases} \tag{4}$$

$$a(v) = \frac{\mu C(v)}{(m_d + m_s)} \tag{5}$$

where μ is the coefficient of action of the resultant force, m_s, m_d represents the static load and dynamic load of the train respectively.

$$E = \sum_{k=0}^{n} C(v_k) \cdot s_k \tag{6}$$

The total energy consumption of the train in the operation section mainly depends on the stress of the train in the sub section. Where k is the number of subintervals, $k \in [0, n]$; s_k is the length of k subinterval.

$$E_{k,stroed} = T_k \cdot P_{f,r} = \frac{1}{2} J\left(\omega_k^2 - \omega_{k-1}^2\right) \tag{7}$$

$$\omega_k^2 = \omega_{k-1}^2 + \frac{2E_{k,stored}}{J} \tag{8}$$

where $E_{k,stored}$ is energy storage of flywheel energy storage device in k subinterval; ω_k, ω_{k-1} represent the flywheel rotation speed in k and (k-1) subinterval; J is the moment of inertia of flywheel, kg·m^2; $P_{f,r}$ is rated power of flywheel energy storage device, w.

The objective function is shown in Eq. 9:

$$E = \min\left(\sum_{k=0}^{n} \mu F(v_k) \cdot l_{gap} - E_{k,stored}\right) \tag{9}$$

3 Solution Approach – DP

Dynamic programming (DP) is a theory first proposed by mathematician Richard Bellman in 1953 to solve multi-stage decision-making problems, rather than a specific algorithm or a specific mathematical model. Dynamic programming can effectively solve the problem with the property of optimality principle, which means that any decision subsequence contained in the decision sequence is always optimal, and satisfying the optimality principle can ensure that the discretized state has no aftereffect. So, the core of dynamic programming is to determine the Bellman equation according to the bellman optimality principle.

The problem can be discretized by evenly dividing the train operation intervals, and some sub intervals can be obtained [11]. Each interval corresponds to a driving state containing speed information, so that the problem of solving continuous speed curve can be transformed into a set of sub interval states. Suppose the length of the train operation section is l, and the operation section is divided into n sub sections. The length of each sub section is recorded as l_{gap}, and the set of all sub sections is recorded as S_k, thus (n + 1) states are divided. The state set corresponding to the subinterval can represent the vehicle speed and flywheel rotated speed at this position, which is shown in Eq. 10 (Fig. 4):

$$\begin{aligned} S_k &= \{s_{k,1}, s_{k,2}, \ldots, s_{k,n}\} \\ &= \{(x_{k,1}, v_{k,1}, \omega_{k,1}), (x_{k,2}, v_{k,2}, \omega_{k,2}), \ldots, (x_{k,n}, v_{k,n}, \omega_{k,n})\} \end{aligned} \tag{10}$$

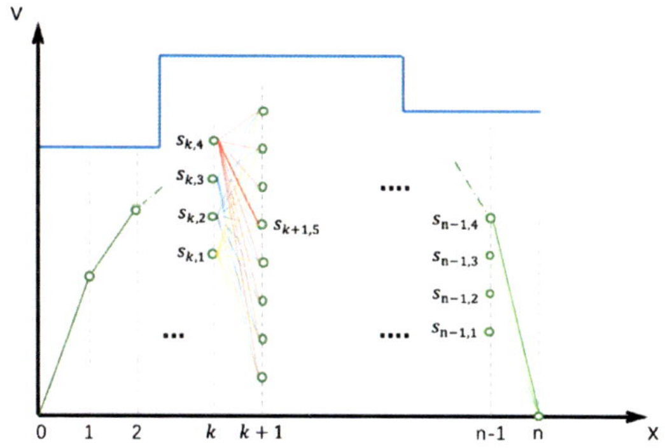

Fig. 4. Schematic diagram of dynamic planning state transfer

By traversing the state transition matrix and analyzing the time and energy consumption of each transition, the optimal utility function can be solved quickly in the process of dynamic planning, which provides the basis for action selection. The objective function of the optimal velocity curve is used to describe the state transition:

$$G(s_{k,v_i}, s_{k+1,v_j}) = \alpha\left(M_E - E_{k+1,j}^{k,i}\right) + \beta\left|T_{t,k} - T_{k+1,j}^{k,i}\right| + \gamma E_{k,stored} \tag{11}$$

where $G(s_{k,v_i}, s_{k+1,v_j})$ is the function value of state transition; α, β and γ respectively represent the weight of energy consumption, time consumption and energy storage in the evaluation of transfer, which are greater than 0 and $\alpha + \beta + \gamma = 1$; M_E represents a constant for energy consumption, and as the reduced number of energy consumption $E_{k+1,i}^{k,1}$ corresponding to state transition, the lower the energy consumption, the better the effect; $T_{t,k}$ represents the reasonable time consumption of k stage in the interval, and the closer the actual time consumption is, the better the effect is.

$$G_k = \begin{bmatrix} G_{k,1} \\ G_{k,2} \\ \vdots \\ G_{k,n} \end{bmatrix} = \begin{bmatrix} G_{k_1}\left(s_{i_1}, s_{j_1}\right) & G_{k_1}\left(s_{i_1}, s_{j_1+1}\right) & \cdots & G_{k_1}\left(s_{i_1}, s_{j_1+m}\right) \\ G_{k_2}\left(s_{i_2}, s_{j_2}\right) & G_{k_2}\left(s_{i_2}, s_{j_2+1}\right) & \cdots & G_{k_2}\left(s_{i_2}, s_{j_2+m}\right) \\ \vdots & \vdots & \ddots & \vdots \\ G_{k_n}\left(s_{i_n}, s_{j_n}\right) & G_{k_n}\left(s_{i_n}, s_{j_n+1}\right) & \cdots & G_{k_n}\left(s_{i_n}, s_{j_n+m}\right) \end{bmatrix} \tag{12}$$

G_k in matrix G represents the result of the effect evaluation function of the transition from state s_i in phase k to s_j in phase k + 1. By this analogy, we can get the effect evaluation matrix of each stage in the whole process of the interval. According to the principle of Bellman optimality, we can get the optimal index function of the inverse solution of dynamic programming:

$$f_k\left(s_{k,v_j}\right) = \min \begin{Bmatrix} G\left(s_{k,v_j}, s_{k+1,v_0}\right) + f_{k+1}^*\left(s_{k+1,v_0}\right) \\ G\left(s_{k,v_j}, s_{k+1,v_1}\right) + f_{k+1}^*\left(s_{k+1,v_1}\right) \\ \vdots \\ G\left(s_{k,v_j}, s_{k+1,v_m}\right) + f_{k+1}^*\left(s_{k+1,v_m}\right) \end{Bmatrix} \tag{13}$$

$$f_k^*\left(s_{k,v_j}\right) = G^*\left(s_{k,v_j}, s_{k+1,v_i}\right) + f_{k+1}^*\left(s_{k+1,v_j}\right) \tag{14}$$

Equation 13 indicates that for the optimal index of state s_{k,v_j} in phase k, it is necessary to traverse all the states that can be converted to s_{k,v_j} in phase k + 1, and calculate the effect of these transformations. In Eq. 14, $f_k^*(s_{k,v_j})$ represents the optimal solution for the transition to s_{k,v_j} state, and $G^*(s_{k,v_j}, s_{k+1,v_i})$ represents thc action corresponding to the optimal transition.

The above method is brought into the calculation example for verification, assuming that the information of operation section is shown in Table 1, and the information of train and vehicle is shown in Table 2:

Table 1. Section line information

Parameter	Symbol (unit)	Value
Line length	L (m)	1000
Maximum speed limit	V_{max}(km/h)	80
Standard operation time	T_t(s)	82
Floating range	Δ(s)	1
Stage gap	l_{gap}(m)	10

Table 2. Basic train information

Parameter	Symbol (unit)	Value
Static mass of train	M_s(t)	360
Dynamic quality of train	M_d(t)	386
Maximum braking acceleration	$a_{b,max}(m/s^2)$	1.1
Maximum traction acceleration	$a_{t,max}(m/s^2)$	1.1
Maximum traction	$F_{t,max}$(kN)	410
Davis Equation factors	A	10.079
	B	0
	C	0.001334

Table 3. Performance parameters of flywheel energy storage device

Parameter	Symbol (unit)	Value
Quality	M_f(t)	2
Minimum rotated speed	ω_{min}(rad/s)	280
Maximum rotated speed	ω_{max}(rad/s)	560
Rated rotated speed	ω_r(rad/s)	442.72
Moment of inertia	J $(kg \cdot m^2)$	180
Rated power	P_r(w)	3*105
Energy storage	E_r(J)	10.58*106

The basic parameters of flywheel mechanism of on-board energy storage device [12, 13] are shown in Table 3.

The optimized train speed curve can be obtained through the calculation as shown in the Fig. 5.

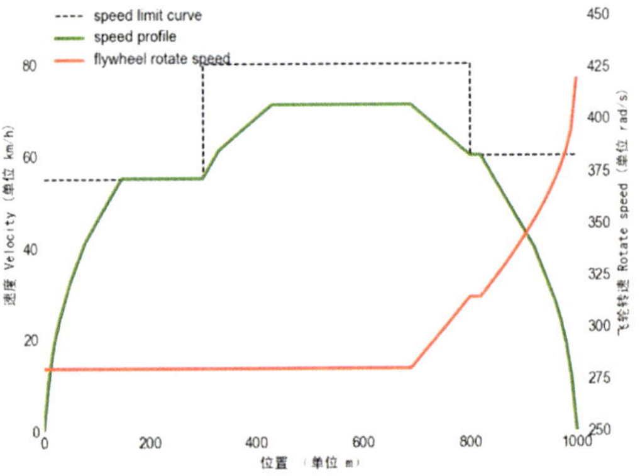

Fig. 5. Train speed curve calculated by dynamic planning

The test results show that the total energy consumption of the train running section is 166.82MJ, which can realize the recovery of the regenerative energy of the train about 8.4 mj, accounting for 5.25% of the total energy consumption. At the end of the section, the SOE of the flywheel energy storage device is 79.22%.

4 Solution Approach– Q-Learning

Reinforcement learning (RL) and deep learning, both of which have developed for many years, are the hot research categories in machine learning. Reinforcement learning is a kind of "environment" and "reward mechanism" established to evaluate the action and realize the optimal control of the decision-making process. The optimal control of urban rail train for the selection of train working conditions and the switching time of working conditions in the operation section conforms to the application direction of reinforcement learning. Q-learning is a kind of reinforcement learning without model. Compared with the model-based method, reinforcement learning without model does not need to bring into the model to solve the decision-making of the action subject [14], so as to plan in advance for the impact of the execution action.

Combined with Q table and R table, the updated Q value of action an in states can be calculated. The update of Q value is based on the Behrman function, just like the effect evaluation function in dynamic programming. The problem solved needs to meet no aftereffect. The Bellman equation on which the Q value is updated is shown in Eq. 15:

$$Q^{new}_{S,A} = Q_{S,A} + \alpha\left(R_{S,A} + \gamma * maxQ\prime(s\prime, a\prime) - Q_{S,A}\right) \tag{15}$$

where $\mathrm{Q}^{\mathrm{new}}_{\mathrm{S,A}}$ is the new Q-value after update; $Q_{S,A}$ is the Q-value in state S after performing action A; $R_{S,A}$ is immediate return on performing action A; $s^{'}$ is the state converted to after performing action A in state S; $maxQ^{'}\left(s^{'}, a^{'}\right)$ represents the maximum Q-value that can be obtained by all executable actions of state S in Q-table; α is learning efficiency, $0 < \alpha < 1$, the greater α is, the greater the weight of the existing Q value is; γ

is discount factor, $0 < \gamma < 1$, the greater the value of γ, the greater the weight of delay return.

To sum up, the pseudo code of Q-learning algorithm using $\in$ greedy search strategy to optimize the solution process of train speed curve can be expressed as follows:

Optimization of train speed curve based on Q-Learning

```
Start learning:
Import the speed curve solved by DP
Set Q-Learning parameters: α、γ、ϵ
Initialize Q-table, set all to 0 by default
Initialize R-table with return function, set -1 in R-table if action can't be implement
For i In Range(train_times):
    Initialize Q-table zero state
    For s In all_states: # States in ergodic intervals
        If sum(Q-table[s]) == 0 Or random.random() > ϵ:
        # if Q-table[s] in initialization or random number is greater
            Randomly select one action implement in Q-value not equal to -1
        Else:
            Select the action with the largest Q-value in the table to execute
        Query R-table, update Q-table
        Initialize state s'
    End For
End For
return Q-table
Forward search Q-table, get the optimized speed curve
End of learning
```

The Q-value updating strategy in the solution process can be expressed as:

$$(s_k, a_i) = (1-\alpha)Q(s_k, a_i) + \alpha\left(R[s_k][a_i] + \gamma * \max\left\{Q\left(s'_{k+1}, a'_j\right)\right\}\right) j, i \in (0, n_a) \tag{16}$$

where $Q(s_k, a_i)$ is Q-value corresponding to a_i of state s_k; $R[s_k][a_i]$ is the return value corresponding action $a_j^{'}$ of state s_k in R-table; $\max\{Q\left(s_{k+1}^{'}, a_j^{'}\right)\}$ is delayed returns; n_a is the number of executable actions.

The calculation of the return function is shown in Eq. 17:

$$R\left(s_{k,a_i}\right) = (M_E - E(v_k, a_i)) + \alpha\left|T_{k,t} - T(v_k, a_i)\right| + \beta\left(T_{k,b} * P_{in}\right) \tag{17}$$

where M_E is the reduced constant of action energy consumption, then the greater the energy consumption and the lower the return; $E(v_k, a_i)$ represents the energy consumption of the action a_i performed in the k stage of the interval; $T_{k,t}$, $T(v_k, a_i)$ represents the time consumption in the k stage when the optimization is not carried out and the time consumption obtained by the execution of the action respectively. $T_{k,b}$ is the duration of brake application; P_{in} is the input power of energy storage device during train braking; α is the time consumption coefficient, which is mainly used to balance the order of magnitude of time consumption and energy consumption; β is the weight of renewable energy.

The main parameters related to Q-learning and the general settings of exploration strategies are shown in Table 4:

Table 4. Q-learning algorithm parameters

Algorithm parameter	Value
Learning rate α	0.1
Discount factor γ	0.9
Greedy coefficient ϵ	0.8
Learning times	3000

The comparison of the optimized speed curve is shown in the Fig. 6:

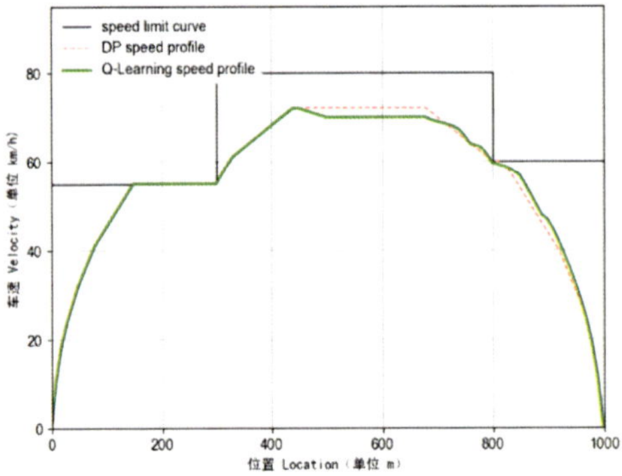

Fig. 6. Train speed curve calculated by Q-learning algorithm

In order to compare the train energy consumption, time consumption and regeneration energy recovery before and after optimization, Table 5 is established for analysis:

Through comparison, it can be found that: when using dynamic programming to solve speed curve, state discretization is based on uniform segmentation of speed interval δ, which helps to solve dynamic programming quickly and simplify state aggregation, and inevitably simplifies the speed after partial state transfer. In the uphill stage of the train (0–400 m interval in this case), if the train chooses the coasting condition, it will react on the speed instead of braking; but if the train chooses the coasting condition in the downhill stage (400–1000 m stage in this case), the speed will change slightly due to the short value of l_gap. In the forward search stage of dynamic planning, it will be

Table 5. Comparison of operation effect before and after Q-learning optimization

Algorithm	Total energy consumption MJ	Time consum s	Brake time consume s	E_{stored} MJ	SOE
DP	184.33	81.81	30.34	9.10	86.0%
Q-Learning	182.04	81.77	32.65	9.80	92.55%

due to the speed integration and conversion in the next stage the former is the same and recorded as cruise condition. This will lead to additional energy consumption of the train, and because the constant speed of IV in the working condition stage makes the speed curve tend to have greater braking force in the braking stage, which is reflected in the braking time of the train will not be conducive to the recovery of braking energy by the flywheel energy storage device.

Because the optimization based on Q-Learning is based on the response of the environment to the action, when establishing the return matrix in the forward search, the operation condition of the train is recorded according to the execution of the action. In this way, the train can use the terrain of the environment in the example to turn to coasting to reduce the running energy consumption, and according to the weight coefficient of the braking time in the call back function calculation, the better collection of the regenerative energy of the train can be realized. From the results in Table 5, it can be seen that through the optimization of Q-learning, the SOE of energy storage device in a single interval can increase the collection of renewable energy by 6.55%, and the total energy consumption in the interval decreases by 2.29mj, accounting for 1.24% of the total energy consumption of the original dynamic planning solution, while the difference between the operation time consumption in the interval and the standard time length is 0.232% and 0.28%, respectively, which are within the allowable error range.

5 Conclusions

Due to the proportion for urban rail transit of power consumption is increasing, so this uses the fly-wheel as the on-board energy storage device to save the braking energy. This method not only can effectively recycle the regenerative energy of the train, but also can reduce the overall energy consumption of the urban rail transit system. Based on the reasonable measure of selecting the fly-wheel as energy storage device, the dynamic planning and Q-learning algorithm is used to optimize train speed curve. The results show that the proposed method can achieve more than 5% energy recovery of the total energy consumption. Therefore, this research has some certain significance for significance for reducing energy consumption of rail transit. However, this research is only at the theoretical and simulation stage, so we need pay more attention to the fly-wheel using as the on-board device. In addition, the train regenerative braking strategy optimization algorithm needs further study.

Acknowledgments. The project is supported by Science and Technology Commission of Shanghai Municipality (20DZ2251900), as well as the Natural Science Foundation of Shanghai

(21ZR1423800). The work is also sponsored and supported by the Key Laboratory of Road and Traffic Engineering, Tongji University and Shanghai Engineering Research Center of Urban Infrastructure Renewal, Shanghai University. The authors are grateful for the reviewer of initial drafts for their helpful comments and suggestions.

References

1. Ichikawa, K.: Application of optimization theory for bounded state variable problems to the operation of train. Bull. JSME **11**(47), 857-865 (1968)
2. Howlett, P.: Existence of an optimal strategy for the control of a train. Sch. Math. Rep. **3** (1988)
3. Oshima, H., Yasunobu, S., Sekino, S.I.: Automatic train operation system based on predictive fuzzy control. In: Proceedings of the International Workshop on Artificial Intelligence for Industrial Applications, 1988. IEEE AI'88. IEEE, pp. 485–489 (1988)
4. Miyatake, M., Ko, H.: Optimization of train speed profile for minimum energy consumption. IEEJ Trans. Electr. Electron. Eng. **5**(3), 263–269 (2010)
5. Domínguez, M., Fernández-Cardador, A., Cucala, A.P., et al.: Multi objective particle swarm optimization algorithm for the design of efficient ATO speed profiles in metro lines. Eng. Appl. Artif. Intell. **29**, 43–53 (2014)
6. Bao, K., Lu, S., Xue, F., et al.: Optimization for train speed trajectory based on Pontryagin's maximum principle. In: International Conference on Intelligent Transportation Systems, pp. 1–6 (2017)
7. Liu, T., Xun, J., Yin, J., et al.: Optimal train control by approximate dynamic programming: comparison of three value function approximation methods. In: 2018 21st International Conference on Intelligent Transportation Systems (ITSC), pp. 2741–2746. IEEE (2018)
8. Wu, C., Lu, S., Xue, F., et al.: Optimization of speed profile and energy interaction at stations for a train vehicle with on-board energy storage device. In: 2018 IEEE Intelligent Vehicles Symposium (IV), pp. 1–6. IEEE (2018)
9. Shen, X.J., Chen, S., Zhang, Y.: Configure methodology of on-board super-capacitor array for recycling regenerative braking energy of URT vehicles. In: Industry Applications Society Meeting. IEEE, pp. 1678–1686 (2012)
10. Radcliffe, P., Wallace, J.S., Shu, L.H.: Stationary applications of energy storage technologies for transit systems. In: 2010 IEEE Electrical Power & Energy Conference, pp. 1–7. IEEE (2010)
11. Yeran, H.G., Lixing, Y., Tao, T., et al.: Train speed profile optimization with on-board energy storage devices: a dynamic programming-based approach. Comput. Ind. Eng. **126**, 149–164 (2018)
12. Gee, A.M., Dunn, R.W.: Analysis of trackside flywheel energy storage in light rail systems. IEEE Trans. Veh. Technol. **64**(9), 3858–3869 (2014)
13. Spiryagin, M., Wolfs, P., Szanto, F., et al.: Application of flywheel energy storage for heavy haul locomotives. Appl. Energy **157**, 607–618 (2015)
14. Lewis, F.L., Liu, D.: Reinforcement learning and approximate dynamic programming for feedback control. IEEE Circuits Syst. Mag. **9**(3), 32–50 (2015)

Design and Implementation of Road Transportation Vocational Skills Service System Based on Mobile Internet

Hong Jia[1,2](✉), Jie Jin[1,2], and Hai-ying Xia[1,2]

[1] Research Institute of Highway Ministry of Transport, Ministry of Transport, Beijing, China
h.jia@rioh.cn

[2] Key Laboratory of Operation Safety Technology on Transport Vehicles, Ministry of Transport, Beijing, China

Abstract. In recent years, with the rapid development of the road transportation industry, the number of road transportation employees has increased sharply. With the concerted promotion of all parties, the safety level of the road transportation industry has been gradually improved, but there are still many potential safety hazards. In order to improve the skill level of employees in the road transportation industry and promote the online vocational skills training of employees in the road transportation industry, starting from the job characteristics and functional requirements of employees in the road transportation industry, this paper designs and implements a road transportation vocational skills service system based on mobile Internet. Based on the jQuery framework, the functions of the system are quickly implemented by using PHP language and other advanced technologies. The system includes PC terminal and mobile terminal, which is convenient for employees in the road transportation industry to study in their spare time at any time. The video course adopts the form of short video, combines theory with examples to facilitate users' understanding, and has achieved good application results. This paper also studies the learning time period of the personnel of automobile inspection institutions in the road transportation industry after the system is put into online application, so as to further optimize the relevant functions of the new curriculum.

Keywords: Mobile Internet · Road Transportation · Vocational skills · Online Learning

1 Introduction

In recent years, the safety level of road transportation industry has been gradually improved, but there are still many hidden dangers. For example, the managers of road transport enterprises are not familiar with the relevant laws and regulations of road transportation, and pay little attention to the continuing education of drivers. The training organized by safety managers of road transport enterprises is not targeted and the effect is poor. The quality of road transport drivers is uneven The driving operation is

X. Zeng et al. (Eds.): ITASC 2022, LNEE 1042, pp. 67–74, 2023.
https://doi.org/10.1007/978-981-99-2252-9_6

not standard and there are many bad habits. Compared with the vocational skills continuing education in Europe and the United States, there is still a certain gap in system and practice in China, and there is still a certain room for improvement. The grass-roots management departments of road transportation also need to improve the professional quality of employees in road transportation industry through an effective way.

Starting from the job characteristics and functional requirements of employees in the road transportation industry, this paper designs and implements a road transportation vocational skill service system based on mobile Internet, which aims to improve the skill level of employees in the road transportation industry and promote the online vocational skill training of employees in the road transportation industry, It also studies the learning time period of personnel in automobile testing institutions in road transportation industry. Based on the jQuery framework, the functions of the system are quickly implemented by using PHP language and other advanced technologies. The system includes PC terminal and mobile terminal, which is convenient for employees in the road transportation industry to study in their spare time at any time. The video course adopts the form of short video, combines theory with examples to facilitate users' understanding, and has achieved good application results.

2 System Requirement Analysis

2.1 Scientific Classification

The number of employees in the road transportation industry is huge, and the knowledge and skills that people in different positions need to master are different. Therefore, it is very important to classify courses and users scientifically. Through the investigation of the road transportation industry, we have carried on the thorough analysis to the system demand, has carried on the scientific classification to the curriculum, the training and the system use organization. Among them, the courses include policies and regulations, standard services, driving skills, emergency response and rescue, and industry services; the training includes four categories: professional qualification training, pre job

Table 1. System related classification properties.

Serial number	Attribute	Content
1	Course classification	Policies and regulations, standard services, driving skills, emergency response and rescue, industry services
2	Training classification	Professional qualification training, pre job training, safety training, continuing education and training
3	Organization classification	Road transport enterprises, automobile production enterprises, vehicle testing institutions, road transport management institutions, emergency rescue enterprises

training, safety training and continuing education; the users of the system include road transport enterprises, automobile production enterprises, automobile testing institutions, road transport management institutions and emergency rescue enterprises. The details are shown in Table 1.

Because of different positions in different institutions, the courses that all kinds of personnel need to learn are also different. Therefore, this paper classifies the target users, is shown in Table 2.

Table 2. System target user classification.

Serial number	Organization classification	User classification
1	Road transport enterprises	Highway passenger transport driver, ordinary freight driver, dangerous freight driver, bus driver, escort, loading and unloading administrator, coach, enterprise leader, safety administrator
2	Automobile manufacturers	Production management personnel, R & D personnel, vehicle type declaration personnel
3	Vehicle inspection organization	Testing management personnel, testing personnel and drivers
4	Road transport management organization	Vehicle management, transportation management, maintenance management
5	Emergency and rescue Enterprises	Maintenance personnel, rescue personnel, management personnel, operators

2.2 Curriculum Design

In order to improve the effect of online learning course, the course has been made scientifically. The popular short video form is used to deepen the learning impression by combining theory with examples. In the system, users should be able to view their own learning progress at any time. For the ranking of learning duration in the organization, they can directly view it to improve their learning enthusiasm.

3 System Design

3.1 System Architecture Design

(1) System technical architecture

The technical architecture of the system is shown in Fig. 1, which is composed of infrastructure layer, data acquisition and transmission layer, data layer, application layer and user layer from bottom to top, as well as information standard specification system, information security guarantee system and information operation and maintenance guarantee system.

The infrastructure layer mainly includes server, storage, operating system, network. The data layer stores all kinds of user information, course information and various documents generated by the system; the application layer includes all kinds of users of the system. The gateway layer is in the way of unified access, flow control, protocol adaptation and other operations to facilitate the call of internal services, current limiting and security. The application layer includes the system PC end and system app end. The user layer is the information exchange channel between the application system and users. The users of the system include various users in road transportation enterprises, automobile production enterprises, automobile detection institutions, road transport management organizations and emergency rescue enterprises, as shown in Table 2.

There are three guarantee systems: the information standard specification system includes the relevant national technical standards that should be observed at all levels in the system implementation. The information security guarantee system provides security support for the system, mainly based on the strict security management system and security technical specifications, to implement the security protection of all levels of the system; the operation and maintenance management guarantee system is the stable and continuous development of the system achievements The important guarantee of the exhibition is to ensure the long-term stable operation and sustainable development of the road transportation vocational skills service system.

The system technical architecture based on mobile Internet is shown in Fig. 1.

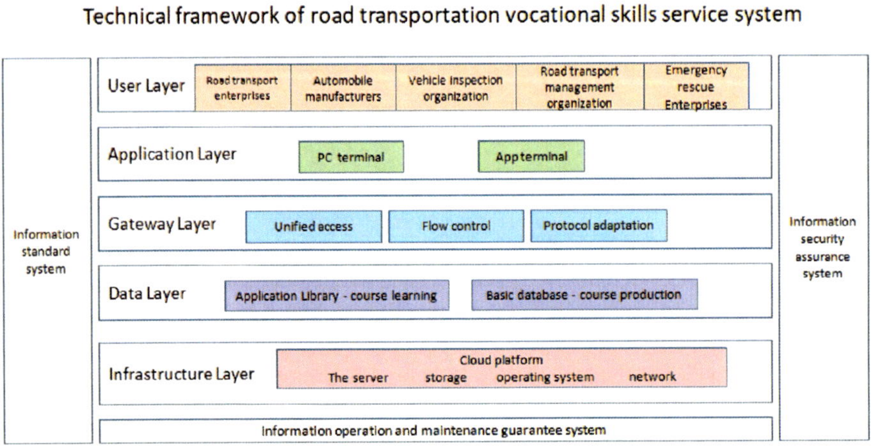

Fig. 1. System technical architecture.

3.2 System Function Design

The system includes PC terminal and app terminal, and the functions are basically the same. The app terminal is designed to be more convenient for users to operate mobile phone based on smart phone. After the organization has completed the registration, the personnel of each post in the organization can log in the system and study the course. Among them, the PC terminal of the system includes the functions of knowledge base, my course, about us and personal center; the app terminal of the system includes home page, knowledge base, learning and my functions. After logging in the system, users can learn post related knowledge and select their favorite courses according to the course classification index. In the system knowledge base, all courses in the current system will be displayed. Under each course classification, the video courses will be sorted according to the learning times of all users. At the same time, each course will show the number of people who are studying. After logging into the system, users can edit their own position information in the "personal center" module, which is convenient for the system to push relevant courses according to their positions. In the "my courses" module, users can learn courses, and can view their learning progress and learning duration ranking in the organization at any time. In the "about us" module, the introduction and contact information of technical support units will be given. If users have any questions, they can leave a message to communicate at any time. The functional architecture design is shown in Fig. 2.

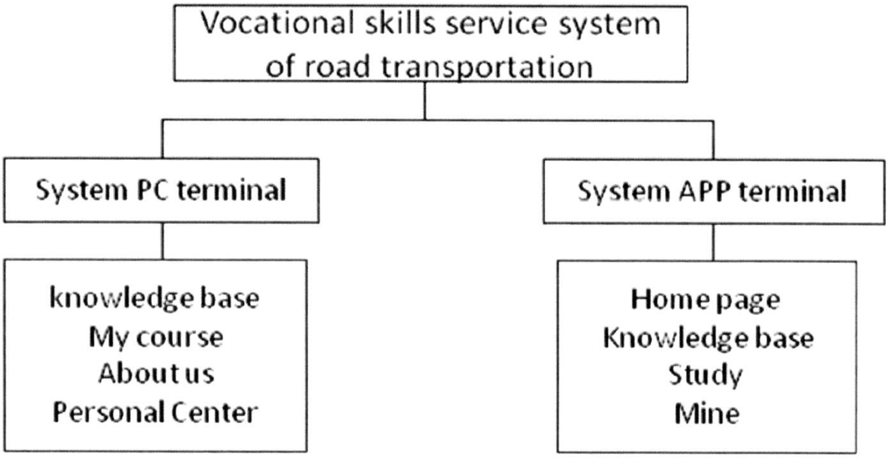

Fig. 2. System function design diagram.

4 System Design System Development and Implementation

4.1 System PC Terminal

The PC terminal of the system is developed based on jQuery framework. The main programming languages are PHP and JavaScript. The CodeIgniter framework is used to implement the multi-dimensional management of courses, and NoSQL and Memcache are used as data storage.

JQuery framework [1–5] is a lightweight JavaScript framework, which greatly simplifies the JavaScript programming, and can efficiently and quickly carry out the front-end system development work. JQuery provides a powerful element selector for getting jQuery objects that encapsulate HTML elements in HTML pages, such as basic selector, hierarchy selector, filter selector and form selector. At the same time, jQuery has built-in some simple and useful built-in animations. It also supports the use of animation based on positioning mechanism to customize animation effects, which provides a lot of convenience for displaying a large number of learning videos in this system. PHP language [6–8] has high development efficiency, supports object-oriented, has good cross platform, can well meet the needs of system development and construction, can well meet the technical requirements of the system in this paper.

4.2 System APP Terminal

The app terminal of the system is developed based on Android architecture. The main programming languages are Java and fluent. The browser with WebKit core is built in. It supports HTML5 web standard, opengles2.0 and lightweight SQLite database.

Java language [9] is a pure object-oriented language with platform independence, good security and robustness. At the same time, Java itself provides a lot of built-in class libraries, which simplifies the programming work of developers and shortens the project development time. Flutter language [10, 11] has a good performance and cross platform language. Together with Java language, it is used as the development language of system app end, which improves the system performance and shortens the development cycle.

4.3 System Deployment

The software system is deployed on two servers on cloud, one as a database server and the other as an application server, which runs well as a whole. The configuration of cloud server is: CPU 4 cores, memory 16 g, operating system Cent OS 7.664 bits, bandwidth 10 Mbps. At the same time, cloud security products such as web application firewall, situation awareness, and knight are configured to ensure the data security of the system.

5 Research on System Application Data

Since the system was put into operation, 999 users have been put into use. After a one-week study period for 100 users of automobile testing institutions, it is found that the number of users who log in to the system on Tuesday is the most; Compared with each time period of the same working day, the number of users who log in to the system from 9:00 to 10:00 in the morning is the largest, followed by 10:00–11:00 in the morning and 16:00–17:00 in the afternoon. The above data show that users of testing institutions have a strong willingness to learn on Tuesday. In each working day, the willingness to learn in the morning is higher than that in the afternoon. Subsequently, we will continue to analyze a large number of data (Table 3).

Table 3. Number of learners of users of automobile testing institutions in each period

Learning period	Monday	Tuesday	Wednesday	Thursday	Friday
09:00–10:00	12	55	42	25	21
10:00–11:00	11	11	15	5	10
11:00–12:00	5	1	6	2	4
12:00–13:00	6	7	5	2	7
13:00–14:00	23	16	3	4	4
14:00–15:00	5	13	9	8	4
15:00–16:00	7	17	8	1	2
16:00–17:00	24	7	11	3	7
17:00–18:00	6	11	5	1	0

6 Concluding Remarks

Based on the full analysis of user function requirements, this paper designs a road transportation vocational skill service system based on mobile Internet, and implements the functions of the system quickly and comprehensively by using advanced technologies such as PHP language based on jQuery framework. The system includes PC terminal and mobile terminal, which is convenient for users to use their spare time for online learning and improve their skills at any time. The video course adopts the form of short video, combines theory with examples to facilitate users' understanding, and has achieved good promotion results. The road transportation vocational skills service system has been put into use, which has effectively promoted the improvement of the skills of employees in the road transportation industry, made them better understand the standards, specifications, policies and regulations, and improved the safety level of the industry. This paper also studies the number of users of vehicle inspection institutions in the road transportation industry in different learning periods after the system is online, so as to further optimize the online related functions of the new course.

Acknowledgments. The project is supported by the scientific and technological innovation fund program of the Research Institute of Highway of the Ministry of Transport (2020-) and the scientific and technological innovation fund program of the Research Institute of Highway of the Ministry of Transport (2018-C0012).

References

1. Xiaode, L.: Design and implementation of power enterprise bidding and supervision system based on jQuery framework. Nanjing University of technology, (2017)
2. Wu, G.: Research and application of rapid development technology of web mobile terminal based on jQuery framework. In: Paper collection of outstanding scientific research achievements of China Vocational Association in 2017 (first and second prize) [C]. Secretariat of China staff education and Vocational Training Association, vol. 6 (2018)
3. Chen, C., Wang, F., Wang, C., Liang, Y.: Design and implementation of OA management system based on jQuery framework. Intern. Combust. Eng. Accessories, **239**(11), 103–105 (2017)
4. Jia, H.F.: Design and Implementation of Science and Technology Consulting Activity Management System based on jQuery Framework. Jilin University (2016)
5. Shen, Y.: Research on web front end development based on jQuery framework. Inf. Commun. **202**(10), 105–107 (2019)
6. Jiang, Y.: Advantage analysis of PHP language in website background construction. Inf. Syst. Eng. **282**(06), 76 (2017)
7. Liu, Y.: Practical analysis of industry website construction based on PHP language. Modern Inf. Technol. **2**(12), 101–102 (2018)
8. Hou, L.: Application of PHP language in enterprise website development. Electr. Technol. Softw. Eng. **142** (20), 2 (2018)
9. Dong, S.: Application of Java programming language in computer software development. Digit. Commun. World **187**(07), 192 + 199 (2020)
10. Li, K., Jia, L., Shi, X.: Design and implementation of SPOC mobile learning platform for medical information technology course based on flutter framework. Comput. Knowl. Technol. **16**(06), 63–64 (2020)
11. Weng, Z., Wu, M.: Design and implementation of image style conversion app based on fluent. Comput. Age **332**(02), 67–70 (2020)

Preliminary Study on Function Layout and Development Principle of Shanghai Soft Soil Deep Underground Engineering Test Base

Qiao Yingjuan[1](✉), Zhu Liangcheng[1], Li Huanqing[2], and Guan Linxing[1]

[1] Shanghai Municipal Engineering Design Institute (Group) Co., Ltd., Shanghai, China
{qiaoyingjuan,zhuliangcheng,guanlinxing}@smedi.com
[2] Associated Research Centers for the Urban Underground Space, Shanghai, China

Abstract. By investigating the industrial function layout of underground engineering test bases under construction or operation at home and abroad, and analyzing the predefined function and industrial layout of deep underground test bases in soft soil area of Shanghai, this article proposes the "1 + 1 + 3" function layout, development and construction principle of deep underground test bases, that is, centered on the deep underground base planning and construction management committee, periodically planning the "2020–2040" strategic layout for integrated construction of deep underground base and taking into account regional planning, hierarchical planning and plane layout of base construction, and provides the construction management guideline for function layout and planning of test bases.

Keywords: deep underground test base · function layout · development planning · construction management guideline

1 Introduction

Underground space construction in Shanghai, with high land use intensity and project diversity, has reached a considerable scale. However, current development of underground space mainly focuses on the shallow and medium covering area within 30 m, and development and utilization of the space has become saturated. Development of deep underground space has become the inevitable choice of urbanization in future. The depth of deep underground space development is mainly 50–100 m. The stratigraphic structure difference and the environmental influence are more complex, which bring a lot of difficulties to construction of deep underground space and selection on the types of operation. Based on the study on deep underground engineering test base, engineering technology utilization and function layout planning of deep underground space are performed to provide theoretical and practical guarantee for the sustainable development of underground space utilization. Currently, with respect to the idea of underground engineering test base, there are cases in China and abroad. Based on the experience of successfully operated test bases, the study on function layout and development procedure of the test base in Shanghai soft soil area is conducted. According to the predefined

X. Zeng et al. (Eds.): ITASC 2022, LNEE 1042, pp. 75–84, 2023.
https://doi.org/10.1007/978-981-99-2252-9_7

function of the test base and the associated industrial development need, the test base development planning and site layout principle are proposed in combination with the surrounding environment.

2 Survey on Industrial Function of Domestic and Foreign Underground

According to the statistics of the underground engineering test bases or scientific laboratories that have been successfully operated or are under construction, the industrial function of test base mainly includes science education, engineering test, logistics & warehousing, infrastructure, and scientific innovation output. Analysis on industrial function of domestic and foreign underground test sites under construction or operation are illustrated in Table 1 (Table 2).

Among these deep underground test sites, Hagerbach Test Gallery, which is successful in engineering test and scientific innovation output, has built a 5.5 km long underground cavity and tunnel system with diverse cross sections over 30 years' development. Engineering tests such as material test, blasting test and tunnel fire control test, and scientific innovation output such as new materials, new machinery (shield machine) and new method (shotcrete technology) are available in test sites. Moreover, test sites also provide venues for study and discussion, dining halls and agricultural R&D tests. Figure 1 shows the inside of Hagerbach Test Gallery.

3 Analysis on Predefined Function of Underground Engineering Test Base

3.1 Analysis on Function Layout of Underground Space

According to the current research, the industrial functions of deep underground space in progress or expected to be necessary in future are as follows:

1) Underground agriculture: The underground space is utilized to develop underground agriculture to realize the crops production mode of high yield, high quality, green and health, and study and develop the technology of plant growth and intervention in artificial sunlight band and culture medium environment.
2) Underground tech lab: Science and technology lab for biomedical science, new material, AI, precise manufacturing and other frontier fields. Underground space can provide enclosed, safe and secure test environment.
3) Underground space base: The moon and Mars are not habitable. In order to explore the universe, humans need to build space bases. Deep underground space can be explored for research and development of life support system technology. The simulated space base can be built for associated technical research.
4) Underground data center: Deploying the critical data center of a city underground will provide a safety barrier against natural disasters such as typhoon and protect data center facilities from all disasters except extreme earthquakes.

Table 1. Analysis on industrial function of domestic and foreign underground test sites under planning, construction or operation

Name of test site/laboratory/project	Industrial function					Industry type
	Engineering test	Scientific innovation output	Science education	Infrastructure	Logistics & warehousing	
Hagerbach Test Gallery, Switzerland	■	■				Industrial, in operation
Kentucky underground data center, USA				■		Industrial, in operation
Sheshan underground data center, Shanghai				■		Industrial, in operation
CTS deep industry base, Switzerland	■	■			■	Industry, under planning and construction
Suzhou River deepwater tunnel, Shanghai	■			■	■	Municipal, under planning and construction
Underground nuclear		■		■		Municipal,

(continued)

Table 1. (*continued*)

installation, Japan						under plannin g and construc tion
Underground neutron energy power station				■		Municip al, under plannin g and construc tion
Underground cemetery				■		Municip al, in operatio n
Josef Underground Research Centre (URC), C		■	■			Scientifi c, in operatio n
Sanford Underground Research Facility (SURF), USA		■	■			Scientifi c, in operatio n
Lantau deep underground physics laboratory, Hongkon			■			Scientifi c, under plannin g and construc tion
Boulby Underground			■			Scientifi c, in

(*continued*)

Table 1. (*continued*)

Laboratory, UK						operation
Underground scientific test site, Singapore		■	■			Scientific, under planning and construction
Space base			■			Strategic, under planning
"Deep sea, deep earth, deep space and deep blue" strategy of China		■				Strategic, under planning and construction

Table 2. 2020–2040 Shanghai integrated construction plan for deep underground engineering test

Construction phasing	Industry type	Industry classification
5-year plan	Industrial	Transportation, data center, robot equipment, new material manufacturing, etc
10-year plan	Municipal	Energy power generation, resource storage, CO2 recycling, livelihood services, etc.
15-year plan	Scientific	Physics experiment, aerospace technology, marine technology, medical technology, etc.
20-year plan	Strategic	Special disaster protection, extreme agriculture, moon habit dwell, etc.

5) Underground cemetery: It satisfies the needs of the deceased for tombs in big cities, and also solves the problem of land scramble between the dead and the living, thus saving land resources. Ecological and future type underground cemetery is built with artificial light, greening and landscape technology to change people's traditional view of catacombs.

Fig. 1. Photos of the inside of Hagerbach Test Gallery

6) Underground energy city: Vigorously developing clean energy, reducing emissions of polluting gases and building energy reserves will become the major trends in maintaining energy safety and promoting sustainable energy development in the future. Deep underground space is used to explore the fields closely related to geoscience, such as geological storage of greenhouse gases and development and utilization of new energy resources such as geothermal energy and NGH. Underground neutron power station is a new type of power station in underground space, which bombards the fuel with high-energy neutrons and provides energy in the form of CHP.

3.2 Analysis on Predefined Function of Shanghai Deep Underground

The function layout of existing underground engineering test bases is investigated and surveyed. At present, according to the development prospect and expected return cycle of different industries, cargo transportation, underground data center, robot, new materials and new technology are the highest in terms of the urgency of demand for deep development of underground space and market. Industries that combine urban development with people's livelihood needs, such as energy power generation, resource storage and solid waste recycling, require early planning and long-term development due to the long planning and construction period and construction planning of at least 10 years will be necessary. Looking at the development of science and technology in today's world, every major breakthrough brings profound economic changes and great social progress. Science and technology has increasingly penetrated into every field of economic construction and human progress and become a decisive factor in the competition in overall national strength. Located deep underground, the underground tech lab is immune from cosmic radiation to a large extent and less affected by atmosphere, temperature and humidity, thus maintaining a relatively stable experimental environment [1]. As a result, global precedents have placed physics, aerospace technology, marine technology and medical technology laboratories underground. Thus, deployment of scientific industries such as medical technology requires construction planning of at least 15 years. Industries with national development strategies such as disaster shelters, extreme agriculture and special equipment require construction planning of at least 20 years.

According to the function layout of the existing underground test base and the demand urgency and planning development law of different industries, the main industrial planning of deep underground engineering test base will be as follows: Industrial, Municipal, Scientific, Strategic.

4 Construction Layout Principle of Deep Underground Engineering Test Base

The function layout of base is planned on the basis of analysis on predefined function of underground test base. Zoning planning, hierarchical planning and plane layout of base are carried out in the field of construction space, and time planning of base is carried out in the field of construction phasing, so as to realize an all-round layout planning of test base in both space and time.

(1) Construction function layout
Zoning planning: The location factors considered in deep underground space development potential research project in Shanghai include thickness of geological layer, industrial potential area, traffic location and vertical function class. Zoning planning is required in the construction distribution planning of deep base. According to the above analysis, in terms of the importance of direct economic effects, development needs of various areas and national science and technology development plans, No. 1 deep underground industrial base, No. 2 deep underground municipal base, No. 3 deep underground scientific base and No. 4 deep underground scientific base are deployed [1]. It is a resource-specific deep underground industry base cluster, as shown in Fig. 2. Hierarchical planning (vertical planning): The functional isolation and structural discontinuity between deep underground space and shallow and medium underground space are beneficial to the independence of deep underground building, including the independence of development sequence and the independence of structural system. In the process of deep underground base planning and construction, vertical distribution planning is needed. Hierarchical planning is helpful for the independent development of the two development modes of special and specialized" deep underground facilities and traditional and popularized" shallow underground facilities, giving consideration to the traditional operation needs of the city and the future-oriented strategic development needs [2]. The functional division of deep base is carried out according to vertical depth, not only meeting the requirements of test functions, but also expanding from single function to different business models to explore the development of a variety of deep underground business models (Fig. 3).

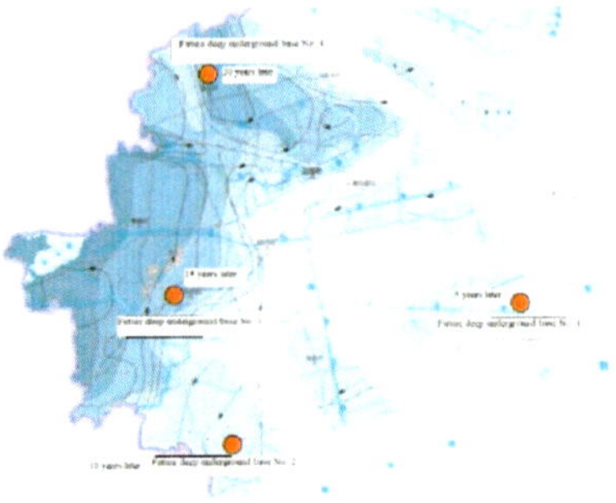

Fig. 2. Zoning planning of deep underground industry base cluster

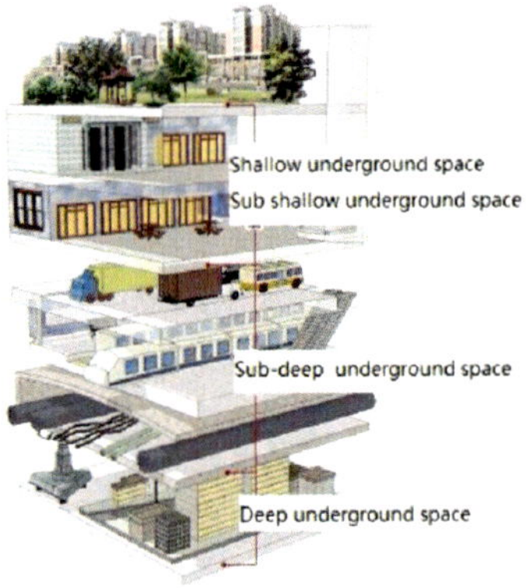

Fig. 3. Planning and utilization of deep and shallow underground spaces [3]

Plane layout: In an industrial base, it is necessary to carry out further plane zoning to improve the planning accuracy. For example, a wedge-shaped greenfield in Shanghai is selected as the site for test base and functional layout and floor planning in the base are carried out. Five shafts are arranged according to the estimated function, requirement and construction method of the test base. In order to maximize the utilization of test function of the site, different construction methods are adopted. For working shafts, the underground continuous wall, pneumatic caisson, open caisson method, reverse building method and VSM method are adopted. For tunnel construction, shield method, pipe-roof freezing and dark cutting, and pipe jacking method are adopted (Fig. 4).

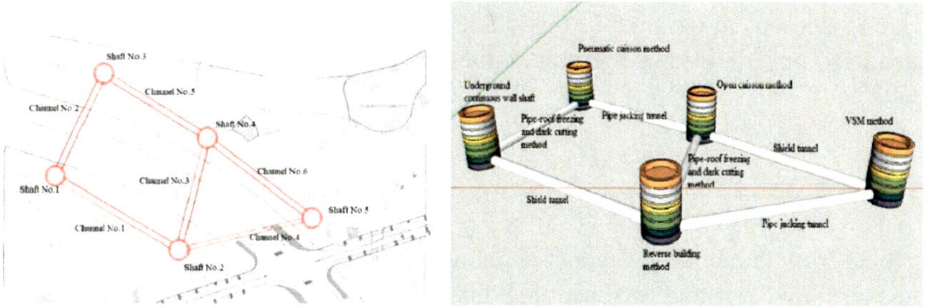

Fig. 4. Plane layout planning of wedge-shaped greenfield deep underground test base

(2) Time planning

From the project development experience of the invested deep underground industrial test bases, the construction phasing of future projects should be given great prominence. Based on the investigation and survey of the administrative department on underground space development in Shanghai, 20 years is a phasing node of urban construction management that fully grasps the administrative system of development and utilization of shallow underground space. The construction of Shanghai deep underground industry base cluster can be divided into four phases (5-year, 10-year, 15-year and 20-year). According to the importance of direct economic effects, the innovation of industry (such as equipment and materials) in deep underground bases is promoted in the first place to meet the most urgent problem in urban development of Shanghai. In addition, align with the world's leading industrial projects to produce ideal market return on investment of the first underground engineering test site. With the construction and management experience of industry, the municipal industries (such as energy and civil affairs) that serve the whole society are developed in deep underground bases for a long time. The scientific industry serves the scientific and technical development of the country, and the strategic industry serves the innovation of human livability, which are long-term construction planning. Based on the resource and asset oriented "construction distribution" planning, the strategic research plan of "2020–2040 Shanghai integrated construction plan for deep underground engineering test base" is proposed, as shown in the following table.

5 Set up "The Deep Underground Base Planning and Construction Management Committee"

A 20-year construction and distribution plan of deep underground engineering test base covering the whole city of Shanghai will lay an important foundation for scientific research innovation and institutional management for the establishment of the Shanghai global base of underground infrastructure. The large-scale economy of deep underground space related industries in Shanghai needs to be built on the basis that the key technical engineering test conditions and the administrative measures have reached certain stage of maturity. Therefore, it is necessary to set up Shanghai soft soil deep underground engineering demonstration area in the whole city, and cooperate with the demonstration area to set up "the deep underground base planning and construction management committee".

6 Conclusion

By learning from the function layout and industrial types of the planned or successfully operated deep underground test bases at home and abroad, the study on function layout and development steps of test base in Shanghai soft soil area is carried out. According to the predefined functions of test base and the development needs of related industries, in consideration of the surrounding environment, the article proposes the view of centering at the deep underground base planning and construction management committee, planning the integrated construction of deep underground test base by phases, taking into account zoning planning, hierarchical planning and plane layout of base construction, and formulating the preliminary "$1 + 1 + 3$" guideline of function layout and development of deep underground test base to provide experience and reference for the planning and construction of Shanghai soft soil deep underground test bases.

This research project is funded by STCSM (STCSM-18DZ1205402)

References

1. Li, H.Q., et al.: Deep Shanghai project – A strategy of infrastructure integration for megacities. Tunn. Undergr. Space Technol. **81**, 547–567 (2018)
2. Li, H.Q., et al.: Deep Shanghai Project – Underground Space Development Potential and Strategy in Shanghai City (a report presented in the 2017 Seminar of Shanghai Municipal Engineering Design Institute (SMEDI), Shanghai, China, 17 October 2017), 136 (2017)
3. https://travel.sohu.com/20090821/n266124378.shtml

Analysis of Industrial Layout Function of Deep Underground Test Base Cluster

Yingjuan Qiao[1](✉), Yiqun Fan[1], and Xurong Ma[2]

[1] Shanghai Municipal Engineering Design Institute (Group) Co., Ltd., Shanghai, China
31615274@qq.com

[2] Zhejiang Communications Resources Investment Co., Ltd., Hangzhou, China

Abstract. According to the strategic research in the "Joint Construction Plan for Deep Underground Engineering Test Bases in Shanghai in 2020–2040", the industrial layout of the four major test bases in the future is proposed. For the main field of technology and situation of development covered by industrial sector, municipal sector, scientific sector, and strategic sector, the analysis is made. Based on the needs of different sectors for development and the sequence of development, the deep underground test ground shall be established to fit in with the directions of the four major sectors, with the vital significance to the research on the four new technologies, the development of the deep municipal facilities, the development of deep resources, and the response to the national "Three-Deep" strategy, on the basis of which the functional layout and planning shall be made.

Keywords: Deep underground test base cluster · industrial layout · industrial sector · municipal sector · scientific sector · strategic sector

1 Preface

The objective of underground space development and exploitation is to meet the needs of sustainable urban development for space and capacity by underground expansion of space, so as to finally achieve the goal of "harmonious city" [1]. The development of shallow underground space in the city has covered underground traffic, municipal utilities, public service facilities, and other functions, and complemented with the functions of surface life, residence, office and recreation, so as to form a spatial system with coordinated operation above and under the ground.

In the "Shanghai Master Plan (2017–2035)", it is pointed out that "For the deep underground space with the scope larger than 50 m, it is required to do a good job in reservation control of the functional systems such as rapid transit, logistics allocation, rainwater storage and regulation, and energy transmission" [2]. Therefore, distinguished from the existing middle-shallow underground space, the facilities suitable for the development of the deep underground space in Shanghai mainly include deep tunnel drainage system, deep underground logistics system, deep underground express road, deep underground storage space, and some underground facilities with specific functions.

X. Zeng et al. (Eds.): ITASC 2022, LNEE 1042, pp. 85–94, 2023.
https://doi.org/10.1007/978-981-99-2252-9_8

Due to the in-depth development of the underground space, the urgent requirements are made for the new technology, new process, new material, and new equipment for control of risks as well as the scientific and reasonable standard system. From the perspective of implementation of national strategy, sustainable urban development, and safeguarding of urban safety height, it is of considerable importance and urgency to establish the deep underground engineering test grounds [3]. The 20-year construction and distribution plan of the deep underground engineering test bases covering the whole territory of Shanghai will lay an essential scientific research innovation and institutional management foundation for establishment of a global underground infrastructure base in Shanghai. According to the strategic research in the "Joint Construction Plan for Deep Underground Engineering Test Bases in Shanghai in 2020–2040", the industrial layout of the four major test bases in the future is proposed. For the main field of technology and situation of development covered by industrial sector, municipal sector, scientific sector, and strategic sector, the analysis is made. Based on the needs of different sectors for development and the sequence of development, the deep underground test ground shall be established to fit in with the directions of the four major sectors, with the vital significance to the research on the four new technologies, the development of the deep municipal facilities, the development of deep resources, and the response to the national "Three-Deep" strategy, on the basis of which the functional layout and planning shall be made (Fig. 1).

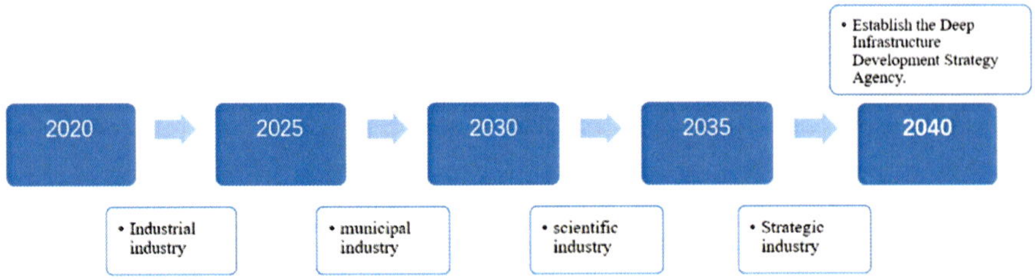

Fig. 1. Deployment of Joint Construction Plan for Deep Underground Engineering Test Bases in Shanghai in 2020–2040

2 Analysis of Industrial Sector

Industrial sector mainly refers to the sector or the engineering of raw material collection and product processing; traditionally, the industry is classified into energy industry, iron and steel industry, machinery industry, hi-tech industry, etc. The industrial sector related to deep underground development covers transportation and equipment manufacturing, data center and hi-tech, robot equipment manufacturing, new materials and their application technologies, and other fields.

By the importance of the direct economic effect, the construction of the deep underground industrial base cluster in Shanghai will first promote the innovation of the industrial sector (e.g., equipment, material) in the deep underground bases, solve the problems in urban development of Shanghai to be solved urgently, and interlink with the most advanced industrial projects in the world, so as to achieve the ideal market economy returns for the investment in the first underground engineering test ground.

Transportation is a major source of carbon emission in the world, and the development of green transportation has become the current trend of transportation development. Underground road and rail transit have become the main direction of urban traffic, but most of them are concentrated in the middle and shallow underground space. With the development of underground space in the future, the development of new transportation in deep underground space, including underground logistics, has gradually become an option. Transportation equipment manufacturing and electrical machinery related to the development of new underground transportation have become potential trades. The research and development of new construction machinery related to underground space construction technology and the research and development of new efficient engineering materials have become the important and urgent problems, because they will have great influence on the construction schedule and construction quality.

Data center is a strategic resource as important as human resources and natural resources. It is the basic physical unit that carries data and becomes the competitive asset of enterprises in the information era. As global climate trend shifts, the probability of extreme weather events increases. According to the investigation report, natural disaster and extreme weather are the two threats at the highest level to the human society, and data center facilities are no exception [4]. Compared with standard ground facilities, the underground data center facilities built in suitable geographical areas are protected from all disasters except extreme earthquakes. Protected by natural geographic conditions and assisted by human security measures, it is possible to ensure all customers' data and hardware will be protected at the top level and to guarantee 100% uptime. With the vigorous development of Internet of Things, the good opportunities are brought for the demand and development of underground data center.

There are many types of underground space operations [5]. Japan, Europe and the United States are the main countries and regions in the world that research and develop the robots for underground operation. The research has also been carried out in the sectors related to underground operation in China, such as the Ministry of Coal, the Ministry of Railways, the Ministry of Water Resources, and the Ministry of Metallurgy. The main varieties researched and developed at home and abroad include shotcreting robot, exploration/detection robot, tunnel boring machine (TBM), disaster relief and rescue robot, mining robot, etc. The development of underground utility tunnel results in the launching of tunnel patrol robot. With the in-depth development of underground space, the storage of robot equipment technology has become another key technology.

3 Analysis of Municipal Sector

Municipal engineering refers to the construction engineering of municipal infrastructures, as well as the construction of infrastructures such as public transport facilities, water supply, drainage, gas, urban flood control, environmental sanitation, and lighting. The industry made up of interrelated municipal sectors with different division of labor related is called municipal sector. With the construction and management experience of the industrial sector, the municipal sector (e.g. energy reserves, water storage facilities, civil affairs facilities) serving the whole society can be developed for a longer period in the deep base.

As a valuable spatial resource, the underground space will definitely become the main source of urban infrastructure construction space in the future, and also gradually become the spatial resource to be seized by various facilities. How to enter the underground space and better perform the service function has become one of the key research fields of power and energy infrastructures [6]. The underground space will play an important role in power grid construction (Cable tunnel, underground substation, and utility tunnel) and distributed energy development (Underground heat storage and geothermal utilization, underground thermal power station, etc.).

With the development of the population and the gradual improvement of living standards, the human demand for energy is increasing. There is also an urgent need for countries to exploit underground energy sources. The deep geothermal energy is the focus of future research and also one of the solutions for the energy crisis. The deep energy drilling project in Iceland is in full swing [7]. In Tengchong of Yunnan, Yangbajing of Tibet, and Qinghai in China, there may be the shallow magma chambers, storing a huge amount of thermal energy. Due to the implementation of the national "deep underground project", such clean energy will also attract great importance (Fig. 2).

Fig. 2. Seeking Energy from Deep Earth (Source of Picture: Courtesy Statoil)

The vigorous development of clean energy, the reduction of pollution emissions, and the setup of energy reserves will become the main trends in maintenance of energy security and promotion of sustainable development of energy in the future. With the deep underground space, the development and exploitation of geological storage of CO_2 greenhouse gas [8], geothermal energy, natural gas hydrate, and other new energies in close relation to the geoscience are explored [9]. The underground neutron power station is a new type of power station which uses the site available from underground space, bombards fuel with high-energy neutrons, and provides energy in the form of combined heat and power. Compared with traditional nuclear power technology, the underground neutron power station has the following advantages: Sufficient fuel reserves, efficient use of fuel, and low cost of electricity (electricity per kWh is less than RMB 0.1 Yuan), underground pre-disposal, redundancy protection of layer, and high safety [10].

In the underground urban ecological conception proposed by Academician Xie [11], the area 50–100 m underground is the underground livable city, the area 100–500 m underground is the underground ecosystem and strategic resource reserve zone, and the area 500–2000 m underground is the underground energy circulation zone, building a strategic blueprint for top-level conception of development at different stages and different levels; this blueprint shows the importance and the locations and functions of the underground energy and the underground water reserves (Fig. 3).

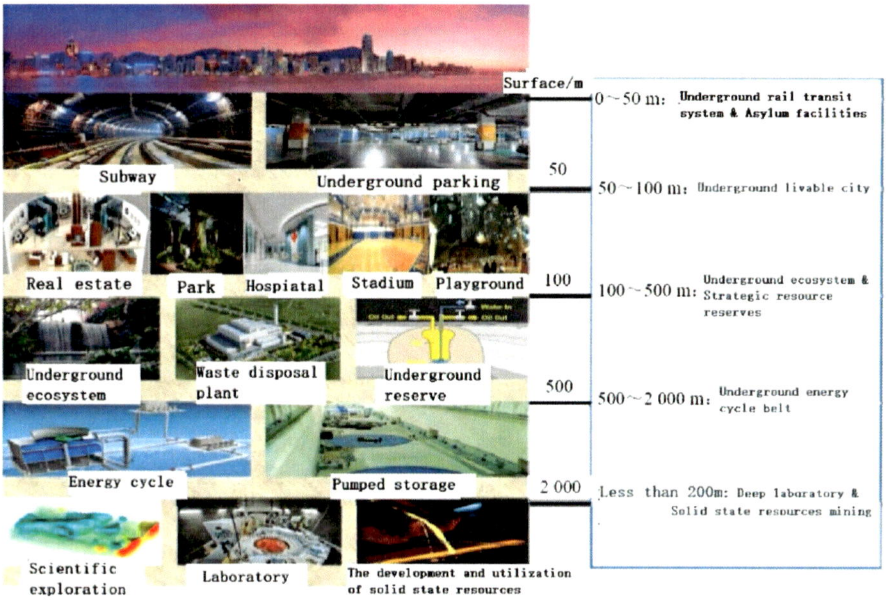

Fig. 3. Underground Urban Ecological Conception [11]

In recent years, the aggravation of the population aging results in the expansion of burial land in China. The consumption of the burial land is becoming more and more serious, which will gradually occupy agricultural land and urban construction land. The underground space of the abandoned mine is huge. When the mining of the whole mine is finished, most of the tunnels, rooms, and unfallen gob areas within the scope of the mine are abandoned deeply underground and become permanent abandoned underground space. The construction of cemeteries with the underground space of abandoned mines can reduce the harm to land resources, water resources and other ecological environments. The underground space of abandoned mines has a broad utilization prospect [12]. Located in Nanjing Niushou Mountain Cultural Tourism Area, the Usnisa Palace is the precedent of large public building in abandoned mine in China [13]. After 21 years of iron ore mining, the massif on the western peak of Niushou Mountain has been leveled off to form a pit with a diameter more than 200 m and a depth more than 60 m. For the purpose of ecological restoration, the Usnisa Palace in Niushou Mountain makes use of the abandoned underground space by layers, combines with history and culture, and fully integrates with the surrounding environment, so as to form the urban underground public space with distinctive tourism features.

In the deep underground space, the deep regulation and storage facilities are laid out for urban waterlogging prevention. There are many cases in the world, such as the Deep Tunnel and Reservoir Project in Chicago USA, the Thames Deep Tunnel Project in the UK, and the Kanda Deep Tunnel Project in Japan. Suzhou River Deep Tunnel Project in Shanghai, as a large deep regulation and storage tunnel laid under Suzhou River, is used for the storage and regulation of urban rainwater capacity to enhance the municipal flood control and drainage capacity. The deepest point of the tunnel project is 60 m underground, with a pipe diameter of 8–10 m. It is planned to set up a full length of 15.3 km, with a storage capacity equal to 400 standard swimming pools. After the completion of the utilities, the flood control standard of the central urban area will be greatly improved. The well-known underground drainage system in Tokyo Japan is located 50 m underground. It is a gigantic tunnel with a total length of 6,300 m and a diameter of 10.6 m. The depth of the pressure regulating tank at the end of the tunnel even reaches 177 m! The whole drainage system makes full use of the function of the underground space and plays a key role in protection of urban area in Tokyo from flood intrusion.

4 Analysis of Scientific Sector

Scientific sector refers to the related industries that take scientific and technological innovation as the guide, focus on industrial technological innovation, carry out research and development of key technologies, and form the transformation of scientific and technological achievements, covering physical experiment, aviation science and technology, marine science and technology, medical science and technology, etc. Scientific sector is a long-term construction plan serving the development of national science and technology.

Singapore Underground Science Town, located in the Singapore Science Park at 80 m–100 m underground as planned, consist of 40 interconnected caves, supporting the data centers and research and development laboratories for development of the biomedical industry. With scientific and technological research and development as an important development direction, Singapore Science Park aims to improve the overall level of national scientific research in Singapore, with research and development expenditure accounting for 2% of GDP (Figs. 4 and 5).

Boulby Underground Technology Test Town is a cutting-edge science and technology laboratory used in biomedicine, new materials, artificial intelligence, precision manufacturing, and other fields. The underground space can provide a confined, safe and secrecy experimental environment.

Extra deep underground science laboratory is not only a basic symbol of national scientific and technological development level, and but also an important direction of underground space development and exploitation [14]. The development of particle physics, astrophysics, cosmology, and other physics expand the field of practical activities of

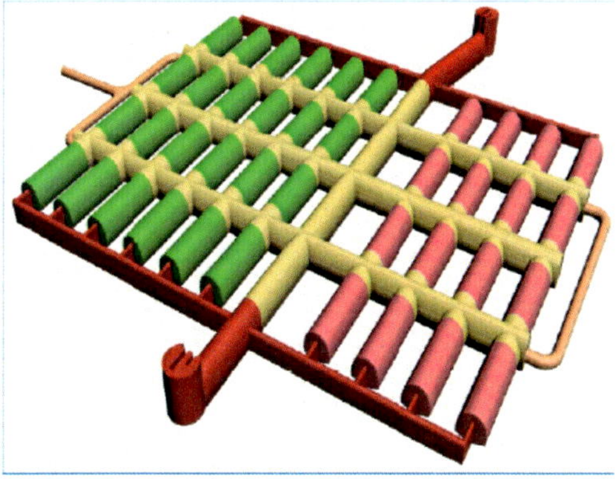

Fig. 4. Singapore Underground Science Town

Fig. 5. Boulby Underground Technology Test Town

human. The underground science laboratory (Hereinafter referred to as underground laboratory), especially the deep underground laboratory is a critical national major basic scientific research facilities; due to the low cosmic ray intensity, low radiation background environment, and other advantages, it becomes the research site of dark matter detection, neutrino physics experiment, and other major cutting edge subjects for particle physics and nuclear physics, astrophysics and cosmology, and other fields. From the perspective of scientific research needs, investment costs, and safety, the functional positioning of the extra deep underground laboratory shall be forward-looking, scientific and diverse, so as to not only meet the experimental needs of different disciplines, but also promote the cooperation of scientific research teams, scientific and technological innovation, and realize the sharing of scientific research resources.

5 Analysis of Strategic Sector

The strategic sector generally refers to the specific industrial sector selected by the state to achieve the high-level objective of industrial structure, which is of great significance to the development of national economy, serve the livable innovation of human beings, and falls within the long-term construction plan. The strategic sector is determined according to different levels of economic and technological development and the prospect of future economic and technological development. At present, the strategic deployment has been made for special disaster prevention, extreme agriculture, and lunar habitation.

China has made it clear that during the 13th Five-Year Plan period, it will promote the three strategies including deep underground exploration, deep sea exploration, and deep space earth observation, strive to become the "leader" in the field of deep underground exploration by 2030, and effectively expand the second prospecting space and underground development space [15]. The future strategic development of deep space is closely related to national economy and people's livelihood.

China is a great country of agricultural production, and the influence of extreme climate factors can hinder the development of Chinese agricultural economy easily [16]. According to statistics, meteorological disasters account for more than 80% of all natural disasters, among which agricultural meteorological disasters are dominant [17]. The long-term equilibrium relationship between the extreme weather and climate events and the agricultural economic output is significant. If the number of days of extreme high temperature, extreme low temperature, extreme precipitation and drought increases by 1%, Chinese agricultural economic output decreases by 0.112%, 0.031%, 0.033% and 0.047% respectively [18]. The underground space is used to conduct underground agricultural research and industrial development, cope with the impact of extreme disasters and urban pollution on agriculture, and make food reserves for the sustainable development of cities.

The underground space has superior functions such as civil air defense, wind prevention, extreme weather prevention, and flood control, etc., and its disaster prevention performance shall be given full play for the intensified urban disasters nowadays [19]. The underground space may be a key battlefield for future wars [20], and play an important role in special disaster prevention.

The surface of the moon and Mars is not suitable for human habitation. In order to explore the universe, humans need to establish space bases, explore and exploit deep underground space, and develop the technology of life support system.

6 Summary

Based on the analysis of the main technical fields and development situation in relation to the industrial sector, municipal sector, scientific sector and strategic sector, the joint construction plan of the four future deep underground engineering test bases is shown as follows:

Industrial sector: 5-year construction plan (transportation, data center, robot equipment, new material manufacturing, etc.)

Municipal sector: 10-year construction plan (energy generation, resource and water storage, carbon dioxide recycling, civil affairs, etc.)

Scientific sector: 15-year construction plan (physical experiment, aviation technology, marine technology, medical technology, etc.)

Strategic sector: 20-year construction plan (special disaster prevention, extreme agriculture, lunar habitation, etc.)

Based on the needs of different sectors for development and the sequence of development, the deep underground test ground shall be established to fit in with the directions of the four major sectors, with the vital significance to the research on the four new technologies, the development of the deep municipal facilities, the development of deep resources, and the response to the national "Three-Deep" strategy.

References

1. Wei, X., Cui, Y.: Discussion on problems and countermeasures for sustainable development and exploitation of urban underground space. Modern City **111**(4), 4–8 (2016)
2. Shanghai Master Plan (2017–2035) Report, p. 43 (2018)
3. Advices on establishment of "comprehensive test base for deep underground engineering". http://www.spcsc.sh.cn/n1939/n3144/n3773/index.html
4. Due to increasingly harsh natural environment, underground data center becomes a trend. http://dc.idcquan.com/ywgl/152708.shtml
5. Su, X., Fan, B.: Underground operation robots and shotcrete robots. Robot. Appl. **3**, 46–49 (2003)
6. Chen, S.: Utilization of underground space in urban electric power and energy engineering. Energy Environ. **4**, 104–105 (2013)
7. Editorial Department of Geology in China: Iceland is looking for energy from underground magma chamber. Geol. China **44**(1), 200 (2017)
8. Underground storage research center of energy & waste. http://www.whrsm.ac.cn/jgsz/kybm/nyyfqwdxccyjzx/
9. Zhang, X., Xiao, H.: Feasibility study on combined utilization of deep geothermal energy and natural gas hydrate. J. Trop. Oceanogr. **34**(2), 85–89 (2015)
10. Finding underground neutron to meet 30% electricity demand in China. http://www.china-nengyuan.com/tech/125361.html
11. Xie, H., Gao, M., Zhang, R., et al.: Strategic conception of underground eco-city and deep underground ecosphere and its key technology prospect. Chin. J. Rock Mech. Eng. **36**(6), 1301–1313 (2017)
12. Lin, X., Zhou, J.: Rational thinking on developing underground space of abandoned mine for cemetery. Nat. Resour. Econ. China **351**(2), 64–68 (2017)
13. Wang, W., Wang, X.: Exploration on renewal and utilization of abandoned underground space in mines to public recreation Space - a case study of Usnisa Palace Project in Nanjing Niushou Mountain Cultural Tourism Area. Jiangsu Constr. **1**, 73–79 (2018)
14. Zhao, R., Yao, Y.: Comparative study on construction of extra deep underground science laboratory in China and abroad. J. Anhui Inst. Archit. Ind. **25**(4), 74–89 (2017)
15. China defines strategic objective for "three-deep" exploration on land, air and sea. http://www.gov.cn/xinwen/2016-09/06/content_5105643.htm
16. Li, C.: Effect of extreme climatic factors on agricultural economy in China. Jiangxi Agric. **16**, 72 (2016)
17. Zeng, X., Guo, X.: Extreme weather, grain yield fluctuation, and agricultural weather risk management. Jiangsu Agric. Sci. **45**(11), 306–309 (2017)
18. Liu, J., Xu, X., Luo, H.: Empirical study on the impact of extreme weather and climate events on agricultural economic output in China. Sci. China: Earth Sci. **42**(7), 1076–1082 (2012)

19. Feng, Y., Cao, Y.: Planning strategy for disaster prevention and mitigation in overall planning of underground space. In: Proceedings of 2014 China Urban Planning Annual Conference (2014)
20. Underground space: The key battlefield of future wars. http://www.81.cn/gfbmap/content/2018-10/18/content_218066.htm

Characteristics Analysis and Prediction of Rail Transit Passenger Flow Based on LSTM

Xiaoqing Zeng[1], Xiaoyuan Yue[1(✉)], Tengfei Yuan[2(✉)], Kaiyi Guo[1], and Dongliang Feng[3]

[1] The Key Laboratory of Road and Traffic Engineering, Ministry of Education, Tongji University, No. 4800 Cao'an Road, Shanghai 201804, China
xy_yue2022@163.com
[2] SHU-UTS SILC Business School, Shanghai University, Shanghai 201800, China
1931341@tongji.edu.cn
[3] Shanghai Municipal Engineering Construction & Development Co., Ltd., Shanghai, China

Abstract. In recent years, many safety accidents caused by high-density passenger flow have occurred in rail transit hub station at home and abroad. The passenger flow prediction of rail transit can not only provide auxiliary decision for managers in passenger flow management and control, but also can provide theoretical guidance for reasonable allocation of traffic facilities around the stations. This paper combs and analyzes the relevant research on rail transit passenger flow prediction at home and abroad, and analyzes its passenger flow characteristics and reasons based on the AFC data of Shanghai rail transit lines. Then, based on long and short-term memory (LSTM) RNN model, the passenger flow prediction of rail transit is realized, and the predicting effects are very good. This method can not only provide data support for urban rail transit operators to make operation plans, but also provide reference for passengers to choose the right travel time and avoid traffic congestion, and can be applied to other rail transit lines.

Keywords: Characteristics Analysis · Passenger Flow Prediction · LSTM Algorithm

1 Introduction

China's rail transit construction is facing unprecedented development opportunities. According to incomplete statistics, the total passenger traffic volume of urban rail transit has reached 23.71 billion in the whole year, an increase of 2.64 billion people times, an increase of 12.5%. With the development of urban rail transit network operation, the passenger volume is expected to increase. Shanghai Metro is an urban rail transit system serving Shanghai and Shanghai metropolitan area in China. It is the longest urban rail transit system in the world. As of January 23, 2021, the maximum daily passenger volume of Shanghai Metro is 13.294 million on March 8, 2019. In 2020, the average daily passenger volume of Shanghai Metro will be 7.7451 million, with a total passenger volume of 2.834 billion.

X. Zeng et al. (Eds.): ITASC 2022, LNEE 1042, pp. 95–104, 2023.
https://doi.org/10.1007/978-981-99-2252-9_9

In recent years, many safety accidents have occurred in the transportation hub stations and public places at home and abroad, which are caused by high density passenger flow [1]. Therefore, accurate prediction of passenger flow development trend becomes an important means to assist managers in passenger flow management and control.

There are a lot of researches on passenger flow forecast of urban rail transit at home and abroad. Chan and others analyzed and predicted the passenger flow of rail transit based on neural network, and improved the prediction model by using mixed exponential smoothing and Ladenburg Marquardt algorithm. The prediction accuracy is high, and it is suitable for small sample data [2]. Jeremy Roos and others proposed a prediction method for passenger flow of Paris urban rail transit. This method uses Kalman filter to filter missing data, and uses multi-dimensional Bayesian distribution to predict historical data. The overall prediction result is better than the previous average and the last observation method, but the model needs to be modified under sudden large flow [3]. Sun Y and others proposed a hybrid wavelet SVM model, which used SVM model to predict passenger flow, and finally reconstructed it through wavelet theory, which improved the prediction performance of the model. Finally, the effectiveness of the model was verified by using Beijing rail transit passenger flow data [4]. Meng Pinchao and others used the moving average model to predict the time series of urban rail transit passenger flow in each period, and compared it with BP neural network, support vector machine and other prediction methods. The prediction results show that the moving average method is better than the above methods [5]. Xie Qiao built a passenger flow prediction model based on nonparametric regression to predict the real-time inbound and outbound passenger flow of rail stations, and analyzed the accuracy of the prediction model by using the historical data of inbound and outbound passenger flow in Guangzhou [6]. Feng Wei and others used GBRBM-DBN algorithm for in-depth prediction of traffic passenger flow, which is suitable for passenger flow prediction under the conditions of time-varying, stochastic and unbalanced [7].

Based on the AFC data of Shanghai rail transit lines from 2018 to 2019, this paper analyzes the characteristics of current passenger flow, and uses LSTM neural network for prediction.

2 Overview of LSTM Model

Long-short term memory is an improved RNN model based on recurrent neural network. It is proposed to solve the problem of gradient dispersion of RNN model. The model adds time series attributes, and is a commonly used model to process time series data in recent years. In the traditional RNN, BPTT is used in the training algorithm. When the time is long, the residual error that needs to be returned will decrease exponentially, resulting in the slow update of network weight, which cannot reflect the long-term memory effect of RNN. Therefore, a storage unit is needed to store the memory, so LSTM model is proposed.

On the basis of ordinary RNN, LSTM adds memory units to each neural unit in the hidden layer, so that the memory information in the time series can be controlled. Each time it passes between each unit in the hidden layer, through several controllable gates (forgetting gate, input gate, candidate gate, output gate), it can control the memory and forgetting degree of the previous information and the current information, so that the RNN network has the function of long-term memory Yes. When an information flows into LSTM network, we can judge whether the information is useful or not by combining rules. Only the information satisfying the algorithm authentication can be left. If it doesn't meet the requirements, the forgetting goalkeeper will choose forgetting.

3 Passenger Flow Forecast Model Based on LSTM

3.1 Construction of LSTM Prediction Model

LSTM time series prediction model is first input layer, then LSTM layer, and then there are two fully connected layers. After dimension reduction, it is connected to one output layer. The input layer is used to receive the continuous sequence input. The length of each input sequence can be set flexibly, but theoretically, if it is defined as 6, 12, 24 and other significant cycles, the effect will be more ideal. This is because the output is also generated by the operation of the input sequence. If the data in the potential period is input, it is more conducive to training. The last layer is the output layer, which outputs the prediction results (Fig. 1).

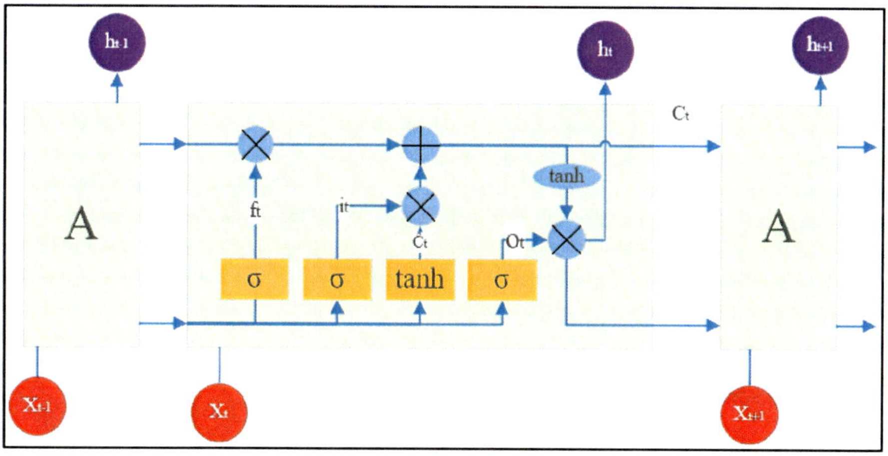

Fig. 1. LSTM network structure diagram

(1) Forget gate

The gate reads h_{t-1} and x_t, output a value between 0 and 1 to each cell in state C_{t-1}. 1 means "fully reserved", 0 means "completely abandoned".

$$f_t = \sigma(W_f \cdot [h_{t-1}, x_t] + b_f) \tag{1}$$

(2) Input gate

The gate determines what value we are going to update. Then, a tanh layer creates a new candidate vector, $\tilde{C}_t$ will be added to the state.

$$i_t = \sigma(W_i \cdot [h_{t-1}, x_t] + b_i) \tag{2}$$

$$\tilde{C}_t = \tanh(W_C \cdot [h_{t-1}, x_t] + b_C) \tag{3}$$

(3) Candidate gate

The gate completes the time update of the old cell state, namely C_{t-1} is updated to $C_{t\circ}$. The old state and f_t times and discard the information we are sure we need to discard. Then add $i_t * \tilde{C}_t$ This is the new candidate, changing according to how much we decide to update each state.

$$C_t = f_t * C_{t-1} + i_t * \tilde{C}_t \tag{4}$$

(4) Output gate

The gate completes the output of the final value. First, we run a sigma layer to determine which part of the cell's state will be exported. Next, we process the cell state through tanh (to get a value between −1 and 1) and multiply it with the output of the sigma gate. Finally, we only output the part of the output that we determine.

$$o_t = \sigma(W_o \cdot [h_{t-1}, x_t] + b_o) \tag{5}$$

$$h_t = o_t * \tanh(C_t) \tag{6}$$

In the above formula, σ is the activation function of the gate, f_t, i_t, o_t is the output of forgetting gate, input gate and output gate respectively, $\tilde{C}_t$ is the activation of cell unit, C_t is the update state of the storage unit, W is the coefficient matrix. Through the function of different gates, LSTM storage unit has better performance than RNN in both short-term and long-term time series prediction.

3.2 Training Model

(1) Autocorrelation analysis

The data of time series are usually analyzed by autocorrelation. Autocorrelation coefficient measures the correlation between the current series value and the past series value, and indicates the most useful past series value when predicting the future value. Through autocorrelation analysis, we can determine the step size of LSTM prediction.

$$\mathrm{ACF}_k = \frac{\sum_{i=1}^{n-k}(x_i - \mu)(x_{i+k} - \mu)}{\sum_{i=1}^{n}(x_i - \mu)^2} \tag{7}$$

where ACF_k is the autocorrelation coefficient of k-order lag, n is the total data volume, K is the lag order, and μ is the total mean value.

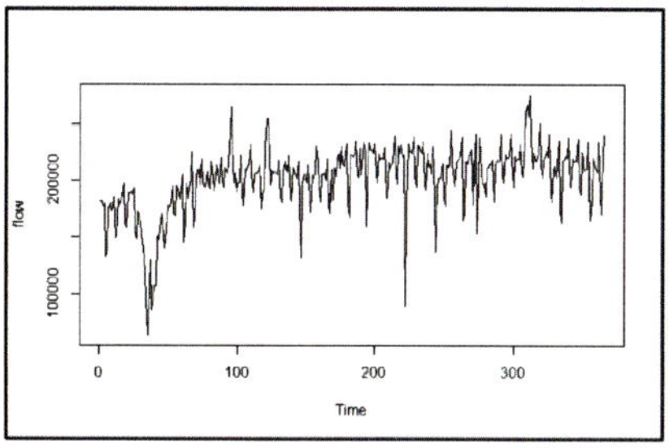

Fig. 2. Sequence diagram

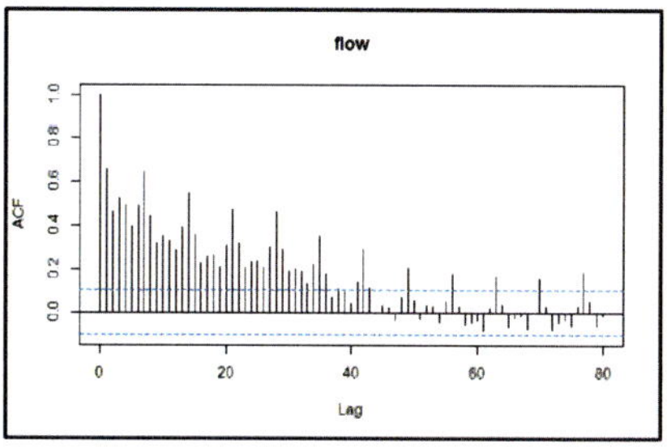

Fig. 3. Autocorrelation diagram

As shown in Fig. 2, the AFC data of Metro has a peak jump about every seven days. As shown in Fig. 3, the horizontal axis lag represents the lag order, the vertical axis represents the correlation coefficient corresponding to each order, and the 0-order lag represents the autocorrelation coefficient of itself, so the corresponding correlation coefficient value is generally 1, and the 95% confidence interval is in the blue dotted line above and below in the figure. That is to say, when the lag order is about 40, the time series converges to 95% confidence interval.

(2) Data partition

The original data set is divided into training set and test set according to a certain proportion. This paper will compare the prediction results of different partition proportions.

(3) Evaluation function

The problems of time series data prediction usually use root mean square error (RMSE), mean square error (MSE), mean absolute percentage error MAPE, mean absolute error (MAE) to evaluate the advantages and disadvantages of a model prediction result. In this paper, the root mean square error is used, which is the square root of the ratio of the square sum of the predicted value to the true value deviation. In the actual prediction, the number of predictions n is always limited, and the true value can only be used to replace the square root error with the most reliable (best) value. Therefore, the root mean square error can reflect the predicted error very well precision.

$$\mathrm{RMSE} = \sqrt{\frac{\sum_{i=1}^{n} (X_{pre,i} - X_{act,i})^2}{n}} \tag{8}$$

4 Case Study

4.1 Characteristics of Full-Day Time-Sharing Passenger Flow

There are obvious differences in the characteristics of hourly passenger flow on weekdays, non-weekdays and holidays of line A, as shown in Fig. 4. There are obvious "double peaks" in the passenger flow on weekdays. The morning peak appears from 7:00 to 9:00, and the hour coefficient of morning peak is 12.4%; the evening peak appears from 18:00

to 19:00, and the hour coefficient of evening peak is 9.5%. The passenger flow in the morning peak is higher than that in the evening peak.

There is no obvious morning and evening peak on non-working days and holidays, and the passenger flow characteristics are similar. The distribution of passenger flow in the whole day is relatively balanced, and the maximum peak hour coefficients are 7.0% and 8.0% respectively.

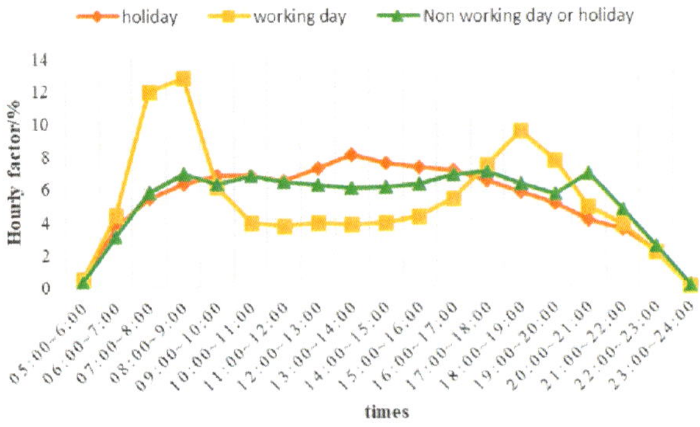

Fig. 4. Characteristics of full-day hourly passenger flow

4.2 Distribution Characteristics of Cross Section Passenger Flow

Weekdays. The passenger flow distribution of the peak section in the early and late working days is shown in Fig. 5 and Fig. 6. The maximum section appears in the upstream of the early peak, and the maximum section passenger flow is 33409 person/h. The overall upstream section passenger flow in early peak is higher than the downward one, and the commuter passenger flow of the above passenger flow is the main one, while the late peak is opposite. The cross-section passenger flow shows the imbalance in direction, and the imbalance coefficients of the early and late peak directions are 1.90 and 2.23 respectively. The maximum passenger flow sections of the upstream and downstream of the early and late peak are all located in the section of Zhuguang road Hongqiao Railway Station. The section passenger flow of Qingpu new town Hongqiao Railway Station is larger in the early peak, which is more than 10000 person/h, and that of Xuying road Hongqiao Railway station under the late peak is larger, and the passenger flow of Zhujiajiao Dongfang green boat section is the lowest, less than 2000 person/h.

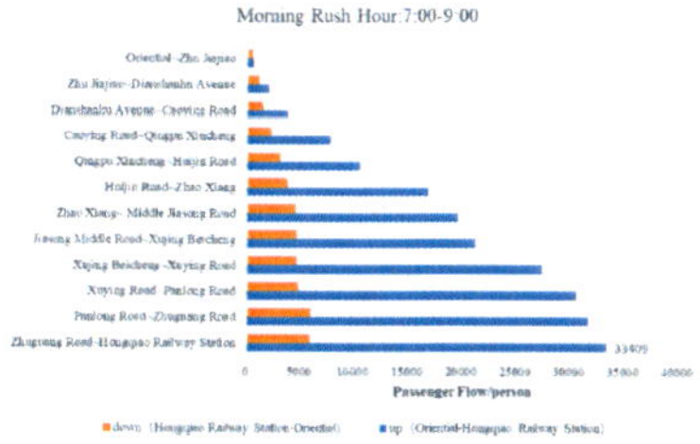

Fig. 5. Passenger flow on weekday morning peak section

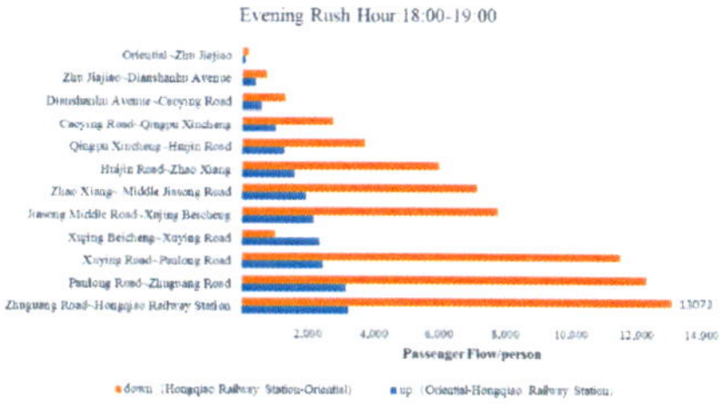

Fig. 6. Passenger flow on weekday evening peak section

Weekends and Holidays. The cross-section passenger flow on weekend and holiday peak hours is generally lower than that on working days, as shown in Fig. 7 and Fig. 8. The up-down passenger flow is slightly balanced compared with that on working days, but the cross-section passenger flow is still unbalanced. The imbalance coefficients of morning and evening peak hours are 1.73 and 1.79 respectively. The location of the section with the largest passenger flow is the same as that on weekdays. The section between Zhuguang road and Hongqiao Railway Station has more than 5000 person/h. The section between Zhujiajiao and Dongfang Lvzhou has the lowest passenger flow, basically less than 1000 person/h.

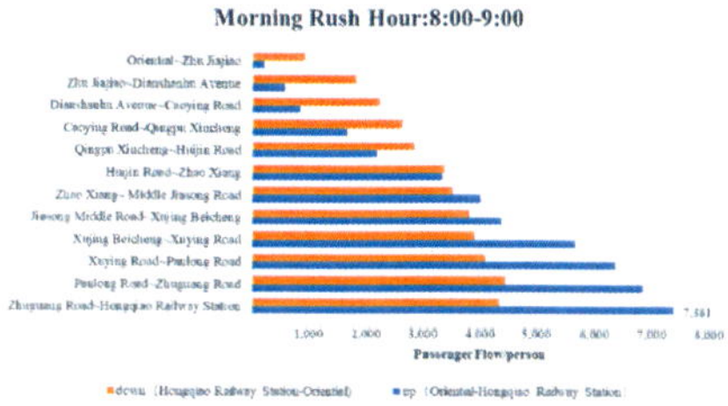

Fig. 7. Passenger flow on weekend and holiday morning peak section

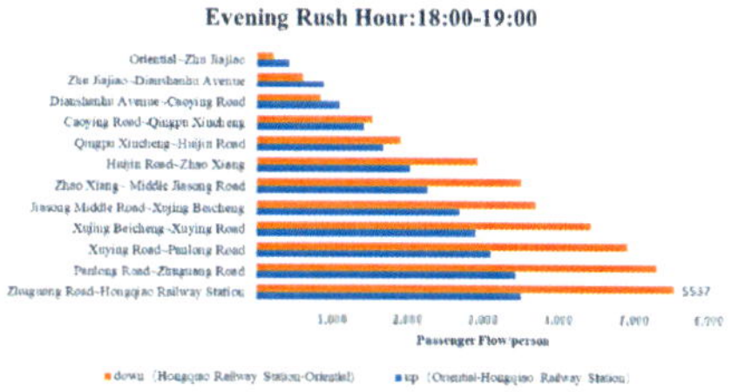

Fig. 8. Passenger flow on weekend and holiday evening peak section

4.3 Characteristics of Daily Transfer Passenger Flow

There is only one transfer station, but the metro entrance is in Hongqiao Railway Station and terminal 2 of Hongqiao Airport. As shown in Fig. 9, The daily average passenger flow of Hongqiao terminal 2 and Hongqiao Railway Station is 5013 and 123000 passengers respectively. The passenger flow of Hongqiao Railway Station is obviously greater than that of terminal 2. The passenger flow of the two transfer stations is basically balanced between line 2 and line 10, but the passenger flow between line 17 and other two lines is obviously unbalanced.

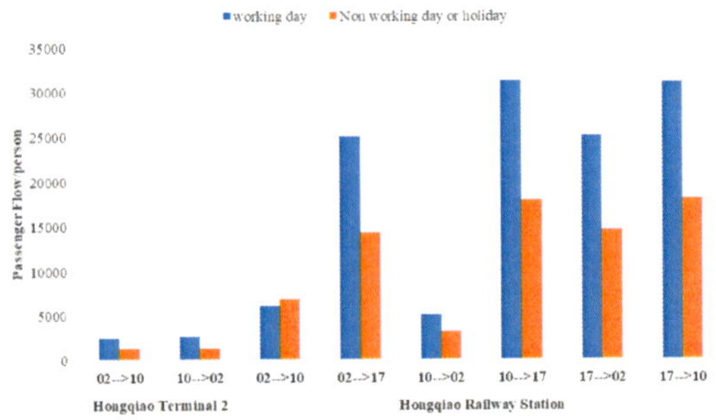

Fig. 9. Average daily transfer passenger flow at transfer station

4.4 Distribution Characteristics of Passenger Flow at Stations

As shown in Fig. 10, Huijin Road, Hongqiao Railway Station and Xujing North City are the stations with larger average daily distribution volume in 2019, among which Huijin road is the station with the largest average daily distribution volume, reaching about 28000 passengers; Dongfang Lvzhou is the station with the smallest distribution volume, with a total daily distribution volume of only about 5800 passengers, accounting for 2.9% of the whole line. In 2018, Hongqiao Railway Station, Huijin road and Xujing North City have the largest daily distribution volume, with about 22000 passengers; Dianshanhu Avenue has the smallest daily distribution volume, with only about 3700 passengers, accounting for 2.3% of the whole line.

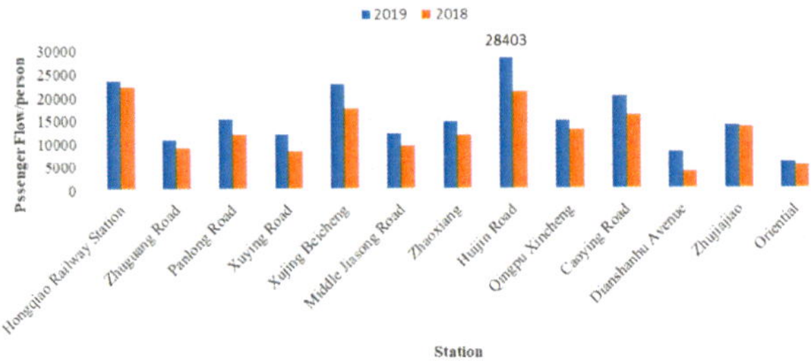

Fig. 10. Passenger flow at each station

4.5 Analysis of Passenger Flow Characteristics

Line A is one of the rail transit lines from the center of Shanghai to the suburbs of Shanghai Qingpu. Since its opening, the characteristics are quite distinct: the passenger flow in peak period is too large, and the average daily passenger flow is too small. According to the characteristics of its passenger flow, it can be seen that the capacity of Shanghai rail transit line A has not been brought into full play. The high passenger flow in peak period actually reflects the passenger flow that should be achieved. The small passenger flow during peak period actually reflects that the line has not been fully utilized.

4.6 Results Analysis of Passenger Flow Forecast Based on LSTM

(1) Parameter setting 1: timestep is 1, train data accounts for 0.67, test data accounts for 0.33. The results showed that the RMSE of train data is 133.08, and the RMSE of test data is 160.73.
(2) Parameter setting 2: timestep is 1, train data accounts for 0.8, test data accounts for 0.2. The results showed that the RMSE of train data is 139.81, and the RMSE of test data is 153.79.
(3) Parameter setting 3: timestep is 7, train data accounts for 0.8, test data accounts for 0.2. The results showed that the RMSE of train data is 139.93, and the RMSE of test data is 157.70 (Fig. 11).

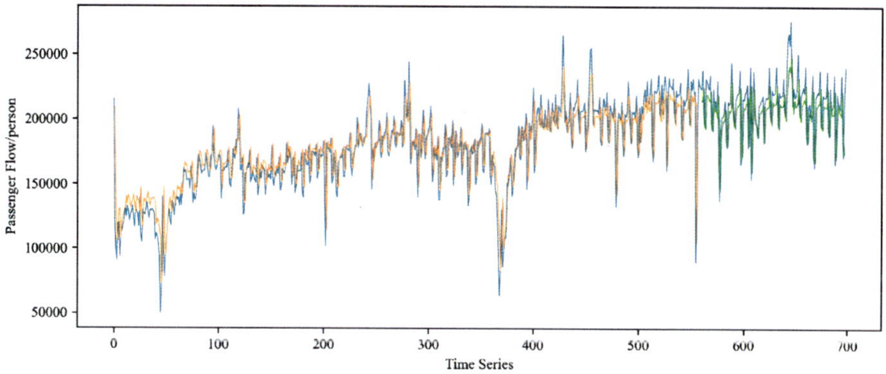

Fig. 11. Results of model training and prediction

From the above three parameter settings, the higher the proportion of training data, the better the prediction effect, which shows that LSTM is very suitable for traffic prediction with previous data basis; in addition, because the time-series correlation of the AFC data is not large, the larger the step size, the lower the accuracy of prediction.

In addition to the date, time, historical passenger flow and other attributes provided in the data set, weather conditions, station environment and other factors will also affect the passenger flow. In the future, data can be supplemented to further optimize the model and reduce the error.

5 Summary

As a suburban rail transit in Shanghai, line A has the characteristics of entering the city in peak period, so its peak section passenger flow is relatively high. On the other hand, the peak capacity of Shanghai rail transit line A has not been fully developed, the root reason is that the purpose of this line is relatively single: Qingpu District peak time into the city, Shanghai International Convention and exhibition center passenger flow. However, with the development of western industry driven by the construction of "Big Hongqiao", the passenger flow of line A shows a significant growth trend.

Based on the AFC data of Shanghai rail transit line from 2018 to 2019, this paper analyzes the current passenger flow of the whole line, including the characteristics of full-time passenger flow, section passenger flow distribution, daily average passenger flow and station passenger flow distribution. Then, the causes of passenger flow characteristics are analyzed. Finally, based on the long and short-term memory LSTM recurrent neural network model, the prediction of the passenger flow of line A is realized, and the minimum root mean square error is about 154, and the forecast effect is good.

The passenger flow prediction of urban rail transit can not only reveal the dynamic trend of passenger flow in and out of the urban rail transit, but also provide data support for the operators of urban rail transit to make operation plan, and dynamically adjust the operation management measures of the train station according to the change trend of passenger flow. In addition, it can also select the correct travel time for passengers in order to avoid traffic congestion, we can strengthen the operation and management of urban rail transit and ensure the healthy operation of urban rail transit. Therefore, the passenger flow prediction of urban rail transit is of great practical significance to both urban rail transit operators and passengers.

Acknowledgments. The project is supported by Shanghai Science and Technology Committee Foundation (Number 19DZ1204202, 20dz1202903-0.1) and Shanghai Municipal Housing and Urban-Rural Construction Management Committee Foundation (Number JS-KY18R022-7).

References

1. Ma, Z.Z., Zeng, X.Q., Huang, J.C., Xiong, Q.P.: Delicacy management system of rail transit passenger origin-destination (OD) flow based on multivariate data fusion. Tunnel Rail Transit (03), 38–41+59–60 (2019)
2. Kit, Y.C., Dillon, T.S., Singh, J., et al.: Neural-network-based models for traffic flow forecasting using a hybrid exponential smoothing and Levenberg-Marquardt algorithm. IEEE Trans. Intell. Transp. Syst. **13**(2), 644–654 (2012)
3. Roos, J., Bonnevay, S., Gavin, G.: Urban rail passenger flow forecasting: a dynamic Bayesian network approach. In: 15th IEEE International Conference on Machine Learning and Applications (ICMLA), pp. 1034–1043 (2016)
4. Sun, Y.X., Leng, B., Guan, W.: A novel wavelet-SVM short-time passenger flow prediction in Beijing subway system. Neurocomputing **166**, 109–121 (2015)
5. Meng, P.C., Li, X.Y., Jia, H.F., Li, Y.Z.: Short-time rail transit passenger flow real-time prediction based on moving average. J. Jilin Univ. (Eng. Technol. Ed.) **48**(02), 448–453 (2018)
6. Xie, Q., Ye, H.X.: Forecast for holiday passenger flow at urban rail transit station based on support vector machine model. Urban Mass Transit **21**(08), 26–29+35 (2018)
7. Feng, W., Chen, H., Zhang, Z.J., Shao, H.P.: A forecast of traffic flow based on GBRBM-DBN model. J. Transp. Inf. Saf. **36**(05), 99–108 (2018)

Traffic Energy Saving Control Based on Reinforcement Learning

Xiaoqing Zeng[1], Kaiyi Guo[1(✉)], Tengfei Yuan[2(✉)], Xiaoyuan Yue[1], Yizeng Wang[3], and Dongliang Feng[4]

[1] The Key Laboratory of Road and Traffic Engineering, Ministry of Education, Tongji University, No. 4800 Cao'an Road, Shanghai 201804, China
Guo_Kaiyi@163.com

[2] SHU-UTS SILC Business School, Shanghai University, Shanghai 201800, China
yuantengfei@shu.edu.cn

[3] Shanghai University, Shanghai, China

[4] Shanghai Municipal Engineering Construction & Development Co., Ltd., Shanghai, China

Abstract. Train energy-saving operation control is a research hotspot in the field of urban rail transit energy-saving. By strengthening the perception ability and decision-making ability of the learning algorithm, this paper puts forward a new idea for the train energy saving control in urban rail transit under the condition of ensuring safety, comfort, real-time and punctuality. To be specific, the following work is done in this paper: (1) Study the related knowledge of train dynamics, establish the train traction model and the train running resistance model and complete the force analysis of the train motion process; (2) Study the knowledge related to energy consumption of train operation and establish the calculation model of energy consumption of trains within the interval; (3) Study the knowledge related to reinforcement learning algorithm, transform the train operation control process into Markov decision process, establish the three elements of reinforcement learning algorithm, and solve the train energy saving control problem by programming. Through simulation, the method proposed in this paper can reduce energy consumption by 13%–17% under the constraints of safety and punctuality.

Keywords: Urban rail transit · Energy saving · Train control · Reinforcement learning

1 Introduction

Compared with other modes of transportation, urban rail transit has the advantages of fast average speed, large passenger capacity, and high utilization rate. However, due to its high operating density and large passenger flow, the total resource consumption value of the urban rail system is also very huge. Among them, train traction energy consumption accounts for more than half. Therefore, studying the energy-saving control of train operation can promote a better green development of urban rail transit.

As shown in Fig. 1, this study minimizes the train operating environment as an optimization goal, takes operating safety, punctuality and other conditions as constraints,

X. Zeng et al. (Eds.): ITASC 2022, LNEE 1042, pp. 105–115, 2023.
https://doi.org/10.1007/978-981-99-2252-9_10

and uses different methods for model transformation. At present, the research methods on such issues are mainly divided into the following categories.

Zhu Jinling through the study of the principle of maximum value, introduced the limited speed constraint condition into the improvement model of the energy-saving driving strategy of the train, and the solution clarified the best control conditions [1]. Liang Zhicheng also used the principle of maximum value to further explore the problem of train handling under restricted speed constraints [2]. Wang Qingyuan used the relevant knowledge of maximum value to introduce the regenerative braking situation of multiple trains into the train control model, and clarified the best set of train control conditions [3]. Although the maximum value method can theoretically obtain the optimal numerical solution, it is precisely because of its too strong theoretical nature that the calculation is complicated.

Wang Pengling used the knowledge of dynamic programming to change the constraint conditions, and combined with the Gaussian pseudo-spectrum method to improve and optimize the train travel speed and distance curve [4].

Shi Hongguo discussed and studied the multi-objective problem of train operation. In order to improve the convergence speed and the performance of the output result, the genetic algorithm was introduced to improve the knowledge of variable length chromosomes, and then the problem was analyzed and solved and improved [5]. Liu Weianalyzed the improvement of multiple population genetic algorithms on the basis of predecessors, and introduced relevant knowledge of variable length real matrix coding, which was used in the improvement of the driving strategy of urban rail trains [6].

As one of the most concerned directions in the field of artificial intelligence in recent years, reinforcement learning provides continuous learning to control the behavior of agents by receiving high-dimensional perceptual input, providing a series of complex strategic decision and perception problems that are currently facing A new approach. Based on the relevant knowledge of the reinforcement learning algorithm, combined with the principle of maximum value, this paper uses the Markov decision-making model of the train running process in the interval, and uses the exploration and learning characteristics of the reinforcement learning to design a single-train energy-saving based on reinforcement learning. The control algorithm is used to solve the improvement strategy of train section driving, so as to obtain the optimal energy-saving strategy.

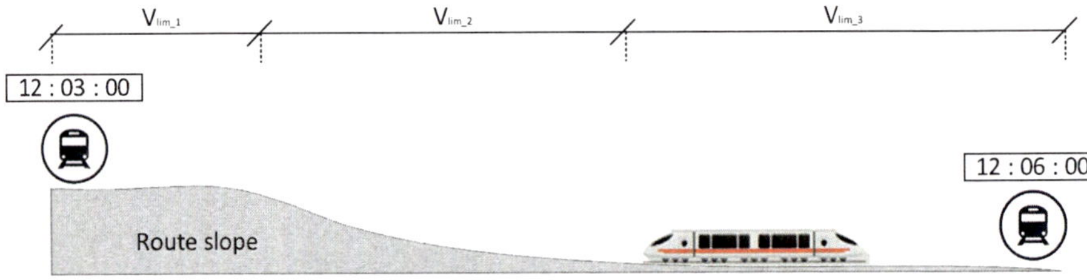

Fig. 1. Schematic scenario for this study

2 Analysis of Train Energy-Saving

When a train is running in a section, it will be affected by external forces such as train traction, train braking force, and train running resistance. Displayed equation is the calculation formula of the resultant external force.

$$F_C = F_T - F_B - F_R \tag{1}$$

FT is train traction (kN), FB is train braking force (kN), FR is train running resistance (kN). When the train runs in the section, it is affected by the traction, braking force, and running resistance of the train. According to the three laws of Newtonian mechanics, displayed equation is construct the kinematic equation of train.

$$\frac{dt}{ds} = \frac{1}{v} \tag{2}$$

$$\frac{dv}{ds} = \frac{\mu_t F_T - \eta_t F_B - F_R}{mv} \tag{3}$$

μ_t is train traction utilization coefficient (ratio of current traction force to maximum traction force, range [0,1]). η_t is rain braking force utilization coefficient (ratio of current braking force to maximum braking force, range [0,1]). The train will generate traction energy consumption during the driving process. This article aims to minimize the energy consumption of train traction, displayed equation is the minimum energy consumption function.

$$minJ = \int_{s_0}^{s_T} [\mu_t F_T(v) - \eta_t F_B(v)]ds \tag{4}$$

When the train is running, it must comply with the corresponding operating rules. Displayed equations are restrictions.

$$\begin{gathered} v(x) \leq v_{Lim}(x) \\ v(0) = 0 \\ v(s) = 0 \\ t(0) = 0 \\ t(s) - t(0) = T_s \end{gathered} \tag{5}$$

V(x) is the speed of the train at position x (km/h). $v_{Lim}(x)$ is the speed limit of the train at position x (km/h). v(0) is the train's running at the beginning Speed (km/h); v(s) is the running speed of the train at the end (km/h). t(0) is the corresponding time at the beginning of the train (s). t(s) is the corresponding time at the ending of the train (s). T_s is the planned running time of the train in the interval (s).

Based on the knowledge of the maximum value, displayed equation is the Hamiltonian function of the train operation strategy.

$$H = -\mu_t F_T(v) + \eta_t F_B(v) + \lambda_1 \frac{\mu_t F_T - \eta_t F_B - F_R}{mv} + \lambda_2 \frac{1}{v} \tag{6}$$

λ_1 and λ_2 are the Lagrange multipliers. Displayed equations are the complementary relaxation factor M(x) to establish a regular equation.

$$\frac{d\lambda_1}{ds} = -\frac{\partial H}{\partial t} \tag{7}$$

$$\frac{d\lambda_2}{ds} = -\frac{\partial H}{\partial t} + \frac{dM(x)}{ds} \tag{8}$$

M(x) meets the constraint conditions of the displayed equations.

$$[v\ \ (x) - vLim\ \ (x)]\frac{dM(x)}{ds} = 0 \tag{9}$$

$$\frac{dM(x)}{ds} \geq 0 \tag{10}$$

Because there is no variable about time t in the Hamiltonian function, λ_1 is a constant. The Eq. (6) can be rewritten as follows equation.

$$H = (\beta - 1)\mu_t F_T(v) + (1 - \beta)\ \eta_t F_B(v) - F_R + \lambda_2 \frac{1}{v} \tag{11}$$

It can be seen from the above equation that if the function J wants to obtain the minimum value, the Hamiltonian function H needs to take the maximum value. Then, since the accompanying variables have multiple values, Table 1 is obtained.

Table 1. Optimal control value of train operation

β	μ_t	η_t
$\beta > 1$	1	0
$\beta = 1$	[0, 1]	[0, 1]
$1 > \beta > 0$	0	1
$\beta = 0$	0	1
$\beta < 0$	0	1

Depending on the value of the accompanying variable, the train operating condition may change (see Table 2). When $\beta > 1$, the value of the control variable corresponds to the maximum traction condition; when $\beta = 1$, the value of the control variable corresponds to any operating condition; when $\beta < 1$, the train corresponds to the maximum braking condition. Therefore, the energy-saving control strategy of train operation. Accelerate with the maximum traction force, then use the cruise mode or idle mode, and use the brake as little as possible during the non-braking operation phase. When the brake is required to stop, the maximum braking force is applied to the train. So, in the following algorithm design in this article, traction and braking are both adopted maximum traction and maximum braking, through the design of the algorithm, find the optimal idle point, in order to achieve as much energy-saving operation as possible.

Table 2. The value of the control variable of the train working condition

Train operating conditions	μ_t	η_t
Traction	(0,1]	0
Cruise	(0, 1)	(0,1)
Coasting	0	0
Brake	0	(0,1]

3 Research and Design of Reinforcement Learning Algorithm

In this paper, combined with the operation control mode of the train in the interval, the Markov decision process is selected as the environment model of the algorithm. Considering the delay time of information acquisition and the amount of calculation and storage of the algorithm, this paper sets the discrete time to 0.2 s. Then, the train is regarded as an agent, the control output of the train in each minute time period is regarded as the behavior set of the agent, and the speed and position of the train in each minute time period are regarded as the current state vector of the train.

Strategy: Strategy refers to the way an agent behaves at a given time or at a given stage, and directly determines the agent's actions or control decisions. In this paper, by modeling the operation process of the train section as a Markov decision process, the output set of the train controller is taken as the action set of the Markov decision process. Then, the action of the agent at stage i is expressed as the displayed equation.

$$u_{\mathrm{i}} = u_{i-1} + \Delta u_i \tag{12}$$

u_i is the train output of the i stage (kN). Δu_i is the train output change of the i stage (kN). Normally, the acceleration of the train in urban rail transit changes between [−1, 1]m/s^2, but for passenger comfort considerations, the maximum traction acceleration of the train in this article is 0.6 m/s^2, and the maximum braking The deceleration is −0.8 m/s^2. Therefore, the change range of the acceleration change rate Δu_i is selected between $[-0.3, 0.3]$m/s^2. In addition, another benefit of reducing the acceleration change rate range is that it can reduce the control variable range to a greater extent, thereby reducing the actual calculation data volume and data buffer volume of the algorithm, and improving the convergence calculation speed of the algorithm.

Reward Function: Reward function refers to the reward value generated by the agent in the process of moving to the next state due to an action taken in the current state. Because this article is to solve the problem of energy-saving operation control of trains, the energy consumption of trains in each small time period is regarded as a reward function. The specific calculation formula is as displayed equation.

$$U_i = \frac{1}{2} * \left| u_0 + \sum_{k=1}^{i} \Delta u_k \right| (v_i + v_{i-1}) \Delta t \tag{13}$$

u_0 is the output of the train starting (kN). v_i is the final speed (m/s) of the i stage of the train. Δt is the time of each discrete time period (0.2 s). Under normal circumstances, the calculation of energy consumption of trains is an integral process of force with respect to distance. But in this article, because the train operation control is broken down into minute time units. Therefore, in a discrete time period, it is approximately considered that the amount of change in force can be ignored, and the calculation of distance can be approximated as a distance calculation method in a straight-line process with uniform acceleration.

Value Function: The value function refers to the weighting and expectation of the reward function, which represents the overall return expectation of the entire control process of the strategy adopted by the agent in the current state. In this paper, the value function is defined as the displayed equation.

$$Q(X, \Delta u) = \frac{1}{k} \sum_{i=1}^{m-1} \gamma U_i \tag{14}$$

γ is discount factor (used to measure immediate repayment and long-term repayment, the range is (0,1]). After defining the state, strategy, reward function and value function above, the value function update formula (15) of reinforcement learning can be derived.

$$Q^{\Pi}(X, \Pi'(X)) = \frac{\varepsilon}{|A(X)|} \sum_{\Delta u} Q^{\Pi}(X, \Delta u) + (1 - \varepsilon) \min_{\Delta u} Q^{\Pi}(X, \Delta u) \tag{15}$$

ε is greedy rate of greedy algorithm. A(X) is the number of action sets in state X.

Table 3 shows the process of the reinforcement learning algorithm to calculate a train speed-distance curve. Because this article needs to solve the problem of energy-saving optimization of train operation control, it is necessary to repeatedly calculate the process to obtain multiple train speed-distance curves, and then make selections to optimize the energy consumption of trains traveling in sections.

4 Simulation and Verification

4.1 Case 1

Case 1 uses the line data of Shanghai Metro Line 17 to verify that the energy-saving algorithm proposed in this paper is effective and reliable in terms of safety, punctuality and energy saving. In this case, the line data from Metro Line 17 from Jiasong Middle Road Station to Xujingbeicheng Station will be used. The running distance between the two stations is 2660 m, the planned train running time of this section is 180 s, and the section information of this section is shown in Table 4.

By importing the relevant line data into the reinforcement learning algorithm for calculation, the most energy-efficient train speed-distance curve in this section is obtained, as shown in Fig. 2.

As shown in the figure, the optimal strategy derived by this algorithm: first perform the maximum traction acceleration, then repeatedly perform the process of idling and

Table 3. Reinforcement learning algorithm calculates energy saving strategy

Step 1: Initially clear all parameters, reset cycle i, speed, distance, status and other information;

Step 2: Control the start of the train with the maximum traction acceleration;

Step 3: Determine the current position of the train $S_i > S - S_d$, where S_d is the train decelerating at the current speed. The required distance. If yes, go to step 9, otherwise go to step 4;

Step 4: Calculate the corresponding reward function U_i according to the state of the train at the last moment and the selected action;

Step 5: Use the greedy algorithm to select the action in the current state, and adjust the output of the train according to the above formula;

Step 6: Determine whether the output selected by the greedy algorithm meets the safety of train operation and other restrictions. If it is satisfied, the output selected by the greedy algorithm is used as the final output; otherwise, the output is adjusted and the output that meets the conditions is used as the final output;

Step 7: Obtain the final output and the state at the next moment, and update the value function according to the value function update formula above;

Step 8: i=i+1, go to step 3;

Step 9: Apply the train with the maximum braking deceleration to make the train stop at the station.

Table 4. Case 1 Line parameter information table

interval/m	slope/%	Speed limit/m
0–200	0	55
200–300	0	80
300–930	– 6	80
930–1800	– 3.081	80
1800–1980	– 3.081	60
1980–2660	0	60

traction, and finally act on the train with the maximum deceleration. And it is not difficult to find that the speed of the train is always below the speed limit of the section during the entire period of the train running. Therefore, this algorithm meets the safety requirements.

Then, according to the actual energy consumption of the train and the running time of the section, it can be obtained from the existing signal system to further enhance the verification of the effectiveness of the algorithm, and refer to the conclusion data [7]. Make a comparison (see Table 5). It can be seen that in terms of travel time, the existing driving strategy is the fastest, followed by this article, and the reference strategy is the slowest. In terms of operating energy consumption, the existing driving strategy is the most energy-consuming, followed by this article, the strategy referred to is the

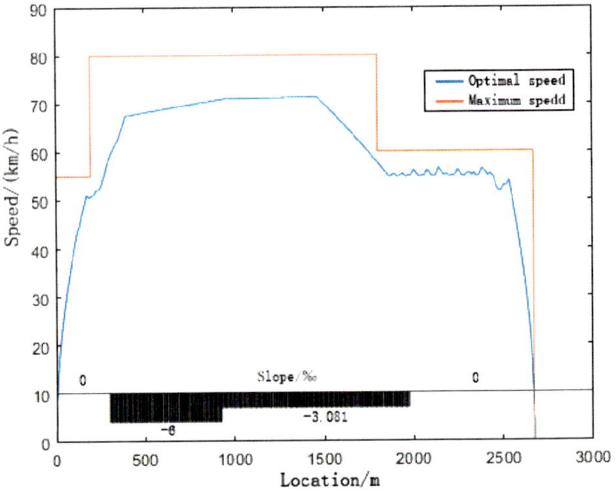

Fig. 2. Optimal control curve diagram based on reinforcement learning algorithm

best. Among them, compared with existing strategies, this strategy can reduce energy consumption by 17.5%. Compared with the energy-saving strategy in the reference, the strategy in this article has a slightly higher energy consumption, but the running time is closer to the standard time. In summary, the algorithm proposed in this paper meets the requirements of safety, punctuality and energy saving in train operation, and has a good energy saving effect.

Table 5. Comparison of three strategies

Driving control strategy	Time/s	Energy consumption/(MJ)
Existing ATO control strategy	179.1	187.21
Reference strategy	181.5	152.44
Strategy proposed in this article	180.7	154.39

4.2 Case 2

In order to further enhance the verification of the effectiveness of the algorithm, Case 2 uses the line data and other information of another reference[7] to compare the output results between different strategies. This case shows that the algorithm in this paper can model in different intervals. Validity in. In this case, the running distance of the section is 1354 m, and the planned running time of the section is 110 s. The slope, speed limit and curve radius information of the section are shown in Table 6.

By importing the relevant line data into the reinforcement learning algorithm for calculation, the most energy-efficient train speed-distance curve in this section is obtained,

Table 6. Case 2 Line parameter information table

interval/m	slope/%	Speed limit/m	Curve radius/m
0–120	0	55	0
120–304	0	80	0
304–684	1.8	80	19
684–1304	– 3.5	80	15
1304–1354	0	80	5

as shown in Fig. 3. It can be seen from the figure that the optimal strategy derived by this algorithm is largely the same as the case: first, the maximum traction acceleration is performed, and then the process of idling and traction is repeated repeatedly, and finally the train is braked at the maximum deceleration. The subtle difference is that the second case starts from the section line 684 m from the downhill section, so the acceleration of the idling here is a positive number, so the train has been idling in this section of the road.

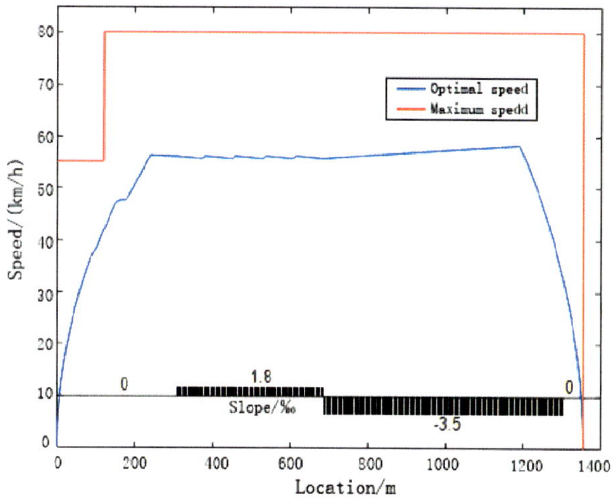

Fig. 3. Optimal control curve diagram based on reinforcement learning algorithm

Then, use the conclusion data in reference [9] to compare (see Table 7). Through comparison, it can be seen that in terms of running time, the existing driving strategy is the slowest, the algorithm in this paper is the second, and the strategy in reference to the literature is the fastest. From the perspective of the error time, the error time of this paper is only 0.2 s, which is 0.6 s lower than the error time of the existing strategy and is within the acceptable range. In terms of operating energy consumption, the existing driving strategy is the most energy-consuming, followed by references, and the strategy in this article is the best. Among them, the strategy in this article can reduce energy consumption by about 13% compared with the existing strategy. Compared with the energy-saving strategy in the reference, the strategy in this article can further reduce the energy consumption by about 8% in this case. In summary, in this case, the algorithm in this paper also meets the requirements of safety, punctuality and energy saving, further verifying the effectiveness and reliability of the algorithm.

Table 7. Comparison of three strategies

Driving control strategy	Time/s	Energy consumption/(kw·h)
Existing strategy for case two	110.8	$3.06*10^7$
Reference strategy	110	$2.89*10^7$
Strategy proposed in this article	110.2	$2.66*10^7$

5 Summary

In this article, for the energy-saving operation of trains, the combination of reinforcement learning algorithm and train operation control is discussed, and the energy-saving optimization model and algorithm of urban rail single train based on reinforcement learning algorithm are proposed. According to the relevant data in the reference, the route and vehicle model are built, and the algorithm designed in this paper is used to solve the problem and verify the validity of the algorithm. Through the comparison of data results, this algorithm has achieved excellent results in both energy consumption and punctual safety.

Although the train energy-saving operation control system based on reinforcement learning designed in this paper has reached good requirements, there are also certain problems worthy of follow-up improvement, such as the establishment of a train multi-particle model.

Acknowledgements. The project is supported by Shanghai Science and Technology Committee Foundation (Number 19DZ1204202, 20dz1202903-0.1) and Shanghai Municipal Housing and Urban-Rural Construction Management Committee Foundation (Number JS-KY18R022-7).

References

1. Zhu, J., Li, H., Wang, Q., et al.: Optimization analysis of train energy saving control. China Railw. Sci. **29**(2), 104–108 (2008)
2. Liang, Z., Wang, Q., Lin, X.: Energy-efficient handling of electric multiple unit based on maximum principle. In: Proceedings of the 33rd Chinese Control Conference, pp. 3415–3422 (2014)
3. Wang, Q., Feng, X., Zhu, J., et al.: Simulation research on energy-saving optimal control of high-speed trains considering the utilization of regenerative braking energy. China Railw. Sci. **36**(1), 96–103 (2015)

4. Wang, P., Goverde, R.M.P.: Multiple-phase train trajectory optimization with signaling and operational constraints. Transp. Res. Part C **69**, 255–275 (2016)
5. Shi, H., Guo, H.: Multi-objective improved genetic algorithm for train operation simulation model. Railw. Transp. Econ. **30**(4), 79–82 (2008)
6. Liu, W., Li, Q., Guo, L., et al.: Research on optimization of energy-saving operation of urban rail trains based on multi-population genetic algorithm. J. Syst. Simul. **22**(04), 921–925 (2010)
7. Liu, L.: Research on Optimal Energy-Efficient Driving Evaluation of Metro Train Based on Flywheel Energy Storage. Tongji University (2020)
8. Miao, C., Wu, S., Zhou, Z., Zhang, W.: Research on energy-saving operation optimization of single train based on time discretization. J. Logist. Eng. Inst. **32**(03), 92–96 (2016)
9. Yin, J.: Research on integrated adjustment method of urban rail train operation based on approximate dynamic programming. Beijing Jiaotong University (2018)

Calculation Method of Switch Machine Health Index Based on Long-Term and Short-Term Neural Network

Tuo Shen[1,2], Zhi Zheng[2], Xiaoqing Zeng[3(✉)], Peiran Ying[3], and Xuanxiong Zhang[1]

[1] School of Optical-Electrical and Computer Engineering, University of Shanghai for Science and Technology, Shanghai 200093, China
[2] Shanghai Key Laboratory of Rail Infrastructure Durability and System Safety, School of Transportation Engineering, Tongji University, Shanghai, China
[3] The Key Laboratory of Road and Traffic Engineering, Ministry of Education, Tongji University, No. 4800 Cao'an Road, Shanghai 201804, China
zengxq@tongji.edu.cn

Abstract. With the continuous development of urban rail, it has become one of the main ways for people to travel. How to ensure the safe and efficient operation of urban rail transit is the focus of attention. The switch machine is one of the important equipment of the signal system. During the long-term service, the performance of the device gradually deteriorates or sudden failure, which will have a significant impact on the operation of the train. For this reason, it is necessary to evaluate the health status of the point machine to predict the life of the point machine. Therefore, a continuously distributed health index (HI) is proposed to characterize the health status of the switch machine in service throughout its life cycle. The operation state of Switch machine is related to the characteristics of its working current. Based on this assumption, this paper monitor the operation state of Switch machine by applying artificial intelligence technology. In this paper, Long-Short Term Memory is trained to predict the health index of the Switch machine. The result shows that the model used by this paper can achieve high accuracy and the prediction accuracy rate reaches 84.9%.It is reliable in predicting the health index of Switch machine. Finally, the HI can be used not only to evaluate the health status of switch machines, but also to predict the health status of other key rail equipment, and promote the development of smart rail transit operation and maintenance.

Keywords: Switch machine · Health Index · Artificial Intelligence · Intelligent Operation and Maintenance

1 Introduction

The operating mileage of urban rail transit in China is growing rapidly. For example, the total operating mileage of subway in Shanghai will reach 700 km in 2020. Railway transportation is one of the key transportation modes for urban passenger flow in Shanghai. By the end of 2020, the operating scale of Shanghai rail transit will exceed

X. Zeng et al. (Eds.): ITASC 2022, LNEE 1042, pp. 116–125, 2023.
https://doi.org/10.1007/978-981-99-2252-9_11

830 km. With the rapid growth of operating mileage, it is necessary to ensure the safety operation of rail transit equipment, which means that facilities and equipment in railway must maintain a highly safe, continuous and reliable operating state throughout their life cycle. The Switch machine is one of the key equipment to ensure the normal operation of the rail transit system. For this reason, it is necessary to monitor the operating status of some important rail equipment, attain the health status of key equipment in real time, therefore make sure the rail transit Operation and maintenance is running safely.

In the field of equipment security, the first proposed study is related to health monitoring [1], and the earliest health monitoring study is about monitoring the health status of fighter jets. With the development of railway transportation, health management technology is getting more and more attention [2]. After health management was proposed, failure prediction technology became popular and many people started to use this technology to conduct related research. Zhen Gao constructed a vehicle health evaluation index system, and established a vehicle health evaluation model based on analytic hierarchy process. It provides scientific help and guidance for the actual maintenance of urban rail transit vehicle equipment units [3]. Lei Bai proposed a new comprehensive evaluation method for the health status of railway track equipment, which can determine the relationship between the multi-state evaluation index of the track grid and the corresponding health index, and finally determine the track grid health index [4].

In 2014, Qian Wang proposed an architecture of the Switch machine failure prediction and health management technology (PHM) based on the failure mechanism of the Switch machine, and constructed and verified the Switch machine power data acquisition system for the S700K electric Switch machine [5]. In 2017, Yuming Fu proposed a Switch machine health assessment method based on fuzzy comprehensive evaluation using various factors of the Switch machine. This evaluation method can better solve the problems of ambiguity and complexity caused by the factors affecting the health of the Switch machine. This method provides a theoretical basis for the implementation of the state maintenance of the Switch machine [6].

Current research on Switch machine is mainly focused on using PHM technology for fault detection and fault diagnosis of the Switch machine, determining the fault type of the Switch machine, and using fuzzy set method to inspect the status Switch machine. However, there are relatively few researches on the quantitative assessment of the health status of the Switch machine, and the evaluation system of the health status has not been established yet. The current operating status of the Switch machine can be attained by analyzing the working current of the Switch machine. Therefore, the deep learning method is used in this article to analyze the operating status of the Switch machine to improve the prediction accuracy. Due to the fact that the working current data of Switch machine is time series data, in this article, we applies the Long-Short Term Memory (LSTM) algorithm to the health evaluation of the Switch machine. This method has high prediction accuracy and can achieve the condition assessment of the Switch machine.

2 Switch Machine Health Index

People use "health" to measure the status of human body in real world, for this we can also apply it to the field of rail transit. In order to know the "health" of the Switch machine, this article use health index to measure the status of the Switch machine. We define the health index as a continuous variable between 0 and 1. The larger the value, the healthier the equipment. For example, when the health index is 0, the device is in a completely faulty state; conversely, when the health index is 1, it indicates that the device is in a completely healthy state. As shown in Fig. 1, on the health index degradation curve, as time goes by, the system health index also shows a downward trend before the system fails. Therefore, the effectiveness of the health index of the Switch machine is essential for the operation and maintenance other equipment and the safe and reliable operation of the track. By using the concept of health index, the operation status of rail transit equipment can be evaluated, and the operation of the equipment or system can be monitored in real time as well.

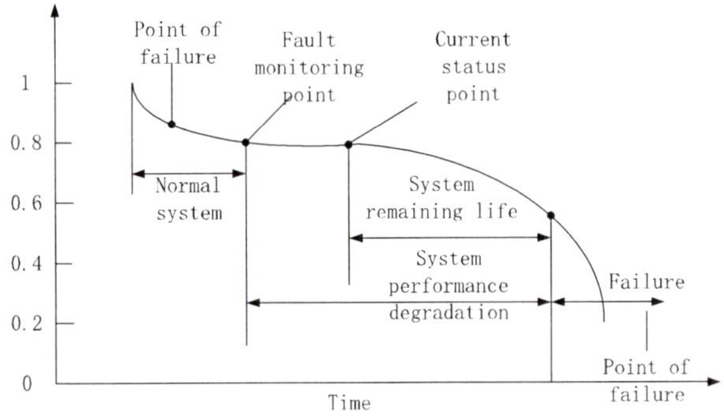

Fig. 1. Health index degradation curve

The operating state of the Switch machine can be induced from its working current curve. Taking the ZD6 Switch machine as an example, the working process of the Switch machine can be divided into the following 4 stages, starting, unlocking, switching, and locking.

The working current curve of ZD6 Switch machine is shown in Fig. 2, it comes from the switch machine equipment of a subway line in Shanghai. In the first stage, when the Switch machine is started, its working current will reach a peak immediately and then drop sharply in a very short period of time. In the second stage, the working current of the Switch machine remains stable after the start-up stage, which means that the unlocking process is completed. When entering the conversion phase, the working current begins to de-crease until the final current value is 0, which means that the Switch machine is locked.

The working current of the Switch machine corresponds to its operating process. If the health status of the Switch machine changes, the corresponding working current will also change accordingly. It can be seen from Fig. 2 that these changes are highly parallel. The operating status of the Switch machine. The working current of the Switch machine has non-linear characteristics, and the neural network has great advantages in solving non-linear problem and can deal with weaker data. Therefore, neural networks can be used to learn from such sample data and discover the hidden regularities.

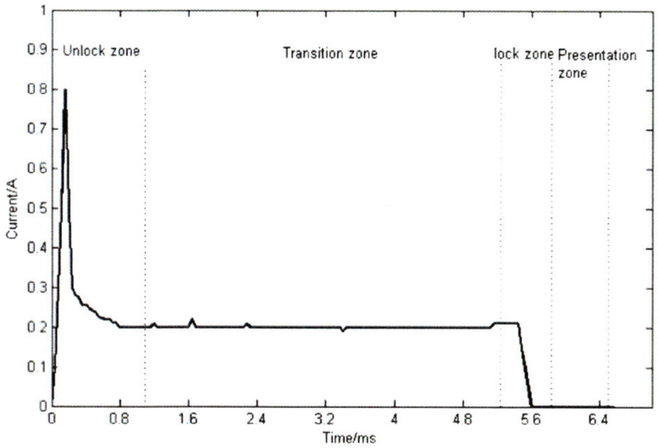

Fig. 2. Switch machine working current curve

For this reason, the real-time data of the Switch machine working process is used as the input data to train our neural network model. The output of the model is the health index of Switch machine, which can be used to assess the health status of the Switch machine. During this process, the real-time fault detection of the Switch machine is achieved by evaluating the operating status of the Switch machine.

3 Research on Calculation Method of Health Index

3.1 Overall Design

The technical route for predicting the health index of the Switch machine is shown in Fig. 3. Firstly, we collect the circuit data of the Switch machine, including the working current data under normal operation and other abnormal situation. Secondly, the data is preprocessed, feature vectors are selected and passed into the LSTM model for training, and the health index of the Switch machine is predicted. At the same time, the divided test set is used for algorithm verification.

3.2 Long Short-Term Memory (LSTM)

The long-short-term memory is a branch of the Recurrent Neural Network (RNN). Its advantage lies in its high efficiency for solving the problem of the gradient disappearance of RNN during the training process. In this article, we build a long short-term memory to train the working current of the Switch machine, which can improve the prediction accuracy.

Unlike RNN, the main feature of LSTM is that it uses a special "gate" structure to determine what information can be transmitted. The structure diagram of LSTM is shown in Fig. 4, which is mainly composed of three gates: input gate, forget gate, and output gate.

Firstly, the model need to filter useful data, which is implemented by passing the data through forget gate. The input value x_t at time t and the output state value h_{t-1} at

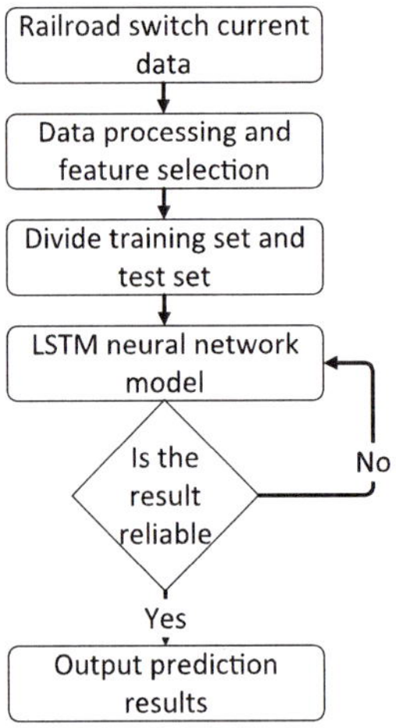

Fig. 3. Flowchart for predicting the health index of the Switch machine

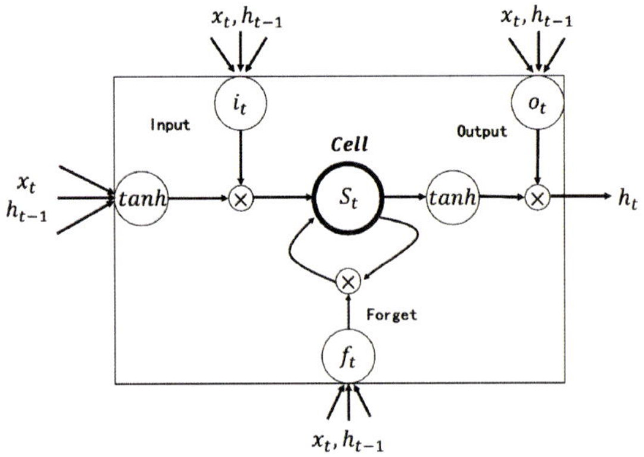

Fig. 4. LSTM structure diagram [7]

time $t - 1$ are passed to the forget gate for data screening, so as to select the useful data for next step. At this time, the output value of cell state C is 1. Conversely, for those data which has little influence for the output, such as abnormal switch current data, the cell state C output value is 0. The labeling of the state is completed through the tanh layer. In this way, the feature data selection of the current data of the Switch machine is completed. The formula of the forget gate is shown in the following formula (1):

$$f_t = relu(W_{f1}x_t + W_{f2}h_{t-1} + b_f) \tag{1}$$

where relu is the activation function; W_{f1}, W_{f2} are the weights of the forget gate; b_f is the offset of the forget gate.

Secondly, input gate and the input node complete the update the cell stat by the input gate and the input node. The formula is shown in formula (2) and formula (3).

Combining formula (1) (2) (3) can generate the current state, as shown in formula (4).

$$i_t = relu(W_{i1}x_t + W_{i2}h_{t-1} + b_i) \quad (2)$$

$$g_t = tanh(W_{g1}x_t + W_{g2}h_{t-1} + b_g) \quad (3)$$

$$s_t = g_t \times i_t + s_{t-1} \times f_t \quad (4)$$

In the formula, W_{i1}, W_{i2} are the weights of the input gates; W_{g1}, W_{g2} are the weights of the input nodes; b_i and b_g are the biases of the input gate and input node.

After the above two steps are processed, the final retained result is calculated by the output gate, and the formula is shown in formula (5). After state update and output gate processing, the output result at this moment is the Switch machine health index, as shown in Eq. (6), and this result is used as the input value at the next moment.

$$o_t = relu(W_{o1}x_t + W_{o2}h_{t-1} + b_o) \quad (5)$$

$$h_t = \tanh(s_t \times o_t) \quad (6)$$

where W_{o1}, W_{o2} are the weights of the output gate; b_o is its bias, h_t is the health index of the Switch machine at time t.

The activation function in the LSTM "gate" structure used in this study is the relu function, which is characterized by faster training and model convergence. It can effectively solving the problems of gradient disappearance and gradient explosion. The formula is as follows:

$$Relu(x) = max(0, x) \quad (7)$$

3.3 Data Normalization

In order to improve the convergence speed and the accuracy of the model, it is necessary to normalize the input data and map the data to the range of 0 to 1 for processing. To this end, the Z-score standardization method is adopted here, as shown below:

$$X = \frac{X - \mu}{\sigma} \quad (8)$$

where μ is the mean value of the original input Switch machine operating current data, and σ is the standard deviation of the original input Switch machine operating current data.

3.4 LSTM Model Parameter Settings

In this paper, we obtain the working current data of the ZD6 Switch machine from the railway station signal centralized monitoring system. The Switch machine is converted and the centralized monitoring system collects 165 sampling points in total with an interval of 40 ms to form a continuous time sequence. In addition to these 165 sets of data as network inputs, the average and maximum values of the working current at each stage of the Switch machine's working process are also used. We then calibrate the data and divide the training set and test set. The number of neurons is set to 50. Initialize the weight of the LSTM layer to a random variable that obeys $U\left(-\frac{1}{\sqrt{50}}, \frac{1}{\sqrt{50}}\right)$, and initialize the weight of the linear transformation layer to a random variable that obeys $U\left(-\frac{1}{\sqrt{50}}, \frac{1}{\sqrt{50}}\right)$. The epoch, batch size and learning is set to 100,100 and 0.005 respectively.

3.5 Example Calculation

3.5.1 Experimental Environment and Data Description

The experimental environment used in this paper is shown in the following Table 1:

Table 1. Table of the experimental environment

Equipment	Model
CPU	AMD Ryzen 5 3550H
RAM	16 G
Programming language	Python 3.8.1
Deep learning framework	Tensorflow

This data set not only collects the current data of the switch machine under normal operation, but also contains data under various fault conditions. The data set division ratio is 8:2, 80% of the original data is the training set, and the remaining 20% is the test set.

3.5.2 The Influence of the Number of Neurons on the Prediction Results

Different numbers of neurons will have different prediction accuracy. This paper uses Mean Square Error (MSE) to measure the prediction accuracy of the model, as shown in formula (9):

$$MSE\left(y^t, y^p\right) = \frac{\sum_{i=1}^{n}\left(y_i^t - y_i^p\right)^2}{n} \tag{9}$$

where y_i^t is the i-th true value, and y_i^p is the i-th predicted value.

Figure 5 shows the relationship between MSE and the number of neurons. The figure shows that there are different MSE under different neurons, and when the number of neurons is 90, the MSE reaches the minimum. For this reason, 90 neurons are selected for training.

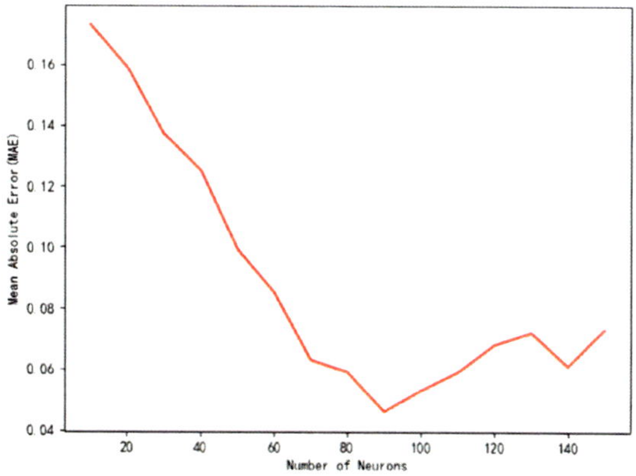

Fig. 5. Relationship between the number of neurons and MSE

After training, the predicted results and the loss function are shown in Fig. 6 and Fig. 7 respectively. The solid line is the true health index, and the dashed line is the predicted health index. Figure 7 shows that the prediction result of the health index is parallel to its real value, which means our model can achieve high accuracy. Table 2 is the Predict HI and Table 3 is the loss of LSTM model.

Table 2. Health Index Predicted Value

Number	HI	HI (Predict)
1	0.9011	0.8592
2	0.8901	0.8584
3	0.8739	0.8421
⋮	⋮	⋮
48	0.8123	0.8045
49	0.8178	0.8071
50	0.8197	0.8095

It can be seen from the prediction results of the LSTM loss function in Fig. 7 that after 100 iterations, the loss function tends to be stable, indicating that the LSTM model has a good fit and is satisfying. After calculation, the prediction accuracy rate reached 84.9% and can be used for health index predictions.

Table 3. Loss of Train and Test

Epoch	MAE of Train	MAE of Test
1	0.1902	0.0556
10	0.0772	0.0248
20	0.0492	0.0167
⋮	⋮	⋮
80	0.0417	0.0152
90	0.0425	0.0163
100	0.0418	0.0155

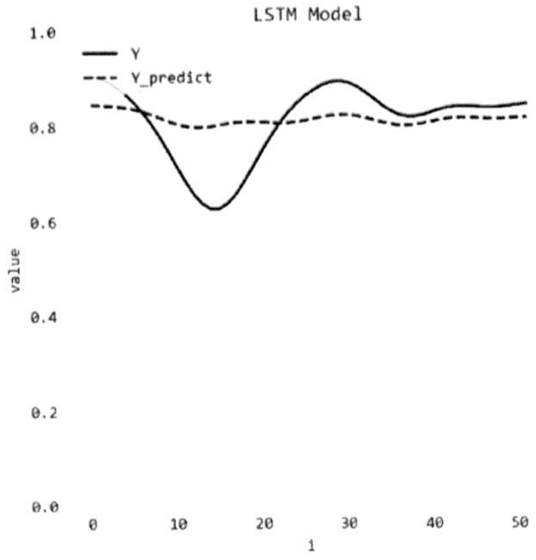

Fig. 6. Comparison chart of the prediction results of the Switch machine health index

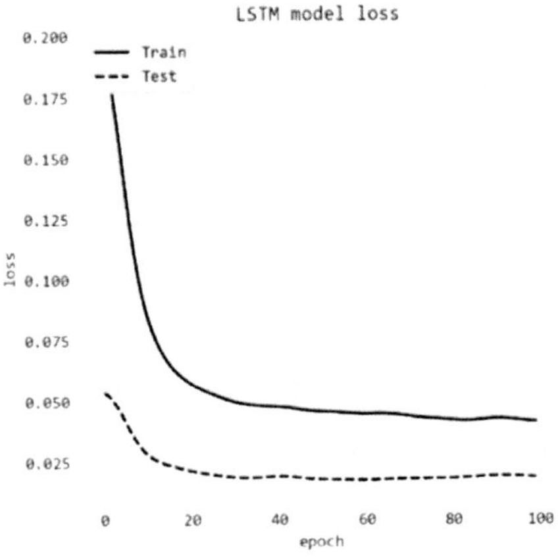

Fig. 7. LSTM loss function calculation result

4 Summary

According to the working process of the Switch machine, this paper proposes a LSTM model to analyze Switch machine's working current data and finally obtain the Switch machine health index. The results show that it can achieve high accuracy and reliability and it's better than simply classifying the Switch machine faults. Secondly, the health index can be predicted in real time based on the real-time detected working current data of the Switch machine and presented through a visual interface, which is convenient for timely maintenance of the Switch machine. It has a leapfrog meaning to transform the traditional fault maintenance to the condition maintenance.

In this paper, we aim to promote smart subway operation and maintenance and the health index prediction model of the Switch machine can also be applied to other rail transit equipment. Fault detection of these equipment can be achieved by predict health index based on their working status. In summary, the application of advanced artificial intelligence technology in the prediction of the health index of rail equipment can achieve high accuracy, thereby help the health management of rail equipment. It also promotes the development of full life cycle track systems.

Acknowledgments. The project is supported by National key R&D projects(2022YFB4300501, Shanghai Science and Technology Committee (20DZ1202900, 19DZ1204200, 23DZ2204900) and Shanghai Housing and Urban-Rural Construction Management Committee (JS-KY18R022-7).

References

1. Hess, A.: The joint strike fighter (JSF) prognostics and health management. In: NDIA Conference. Proceeding of 4th Annual System Engineering Conference, pp. 1–10. JSF Program Office (2001)
2. Larder, B., Azzam, H., Trammel, C., et al.: Smith Industries HUMS: changing the M from monitoring to management. In: Aerospace Conference. IEEE (2000)
3. Gao, Z.: Research of health status assessment for urban rail transit vehicles. Urban Mass Transit **22**(S2), 32–35 (2019)
4. Lei, B.: Grid-Based Decision-Making Models for Railway Track Health Management. Beijing Jiaotong University (2017)
5. Wang, Q.: Construction and Verification of Point Machines Data Acquisition System Oriented to PHM Technology Application. Jinan University (2014)
6. Fu, Y., Liu, B., Song, S.: Research on health assessment methods for switch machine based on fuzzy comprehensive evaluation. J. Railw. Sci. Eng. (05), 192–198 (2017)
7. Zhang, Y., Xiong, R., He, H., et al.: Long short-term memory recurrent neural network for remaining useful life prediction of lithium-ion batteries. IEEE Trans. Veh. Technol. 1 (2018)

Tolerable Hazard Rate Allocation for Urban Rail Automatic Train Control System

Xiaoqing Zeng[1], Yungen Fang[1], and Tengfei Yuan[2,3](✉)

[1] The Key Laboratory of Road and Traffic Engineering, Ministry of Education, China, School of Transportation Engineering, Tongji University, 4800 Cao'an Road, Shanghai, China

[2] SHU-UTS SILC Business School, Shanghai University, Shanghai 201800, China
yuantengfei@shu.edu.cn

[3] Shanghai Engineering Research Center of Urban Infrastructure Renewal, Shanghai 200032, China

Abstract. To improve the reliability and safety of urban rail Automatic Train Control System, it is Query ID="Q1" Text="As Per Springer style, both city and country names must be present in the affiliations. Accordingly, we have inserted the country name in the affiliation 2. Please check and confirm if the inserted country name is correct. If not, please provide us with the correct country name." necessary to allocate the tolerable hazard rate (THR) of the Automatic Train Control (ATC) system to each subsystems and components during the design and implement of specific urban rail transit signaling system. In the beginning, this research analyzes the architecture of ATC system, as well as safety functions and overall safety requirements for the urban rail transit signaling system. Next the specific requirements and principles of equal apportionment technique, safety impact-based apportionment technique and complexity-based apportionment technique are discussed. Combined with the ATC system architecture, safety logic model and specific engineering design parameters, theses three safety allocation methods are used to allocate the THR to each subsystems of ATC as the requirements. At last, the comparison of allocation for these techniques shows that the allocation method based on system complexity is the most suitable for the actual conditions of the project, which can meet the requirements of the Urban Rail Automatic Train Control System. Therefore, this research is proved to be meaningful to improve the reliability and safety of urban rail transit to some extent.

Keywords: Automatic Train Control (ATC) · Tolerable Hazard Rate (THR) · Allocation method · Safety Requirement

1 Introduction

For the modern urban rail transit systems, in order to ensure safe operation of trains, Automatic Train Control (ATC) systems are used to implement the necessary safety functions such as train overspeed protection, maintaining train running intervals, and preventing train collisions. During the construction of an engineering projects, the rail transit systems owner or the operation manager will set a top-level safety requirement

X. Zeng et al. (Eds.): ITASC 2022, LNEE 1042, pp. 126–135, 2023.
https://doi.org/10.1007/978-981-99-2252-9_12

for ATC system, a definitive Tolerable Hazard Rate (THR) value, and requires the final delivered ATC system to meet this targeted safety requirements. At current project practice, the typical urban rail transit Automatic Train Control system is a geographically distributed system, which distribute on the train, track, depot and operation control center, these distributed equipment in different locations cooperate with each other and work together to implement the function of train operation safety control.

However, the distributed Automatic Train Control system devices often come from different equipment suppliers or design institute. During the design and manufacturing process, only the safety requirements of a single device, such as Safety Integrity Level (SIL), Tolerable Hazard Rate (THR) and Tolerable Functional Failure Rate (TFFR) are concerned by the equipment suppliers or design institute. When integrating these devices into the train control system of an urban rail transit line without any unified safety requirements planning, allocation and coordination, the system safety performance achieved by the overall ATC system may not be able to meet the safety requirements set initially. In order to ensure that the final deliverable of overall ATC system of the rail transit line can meet the established safety requirements, the THR need to be allocated during the plan and design stages to guide the design and manufacture activities of individual equipment, which is necessary to ensure the project delivery meet the ATC system operational safety requirements.

2 ATC System Architecture and Safety Requirements

In general, the typical ATC system in current urban rail transit project adopts the communication-based train control (CBTC) system. The architecture and safety requirements of the CBTC system are analyzed as follows.

2.1 CBTC System Architecture and Safety Requirement

According to the definition from IEEE [1], the CBTC system is a continuous ATC system utilizing high-resolution train location determination, independent of track circuits; continuous, high capacity, bidirectional train-to-wayside data communications; and train-borne and wayside processors capable of implementing vital functions. The architecture with its subsystem location is shown in the Fig. 1.

In CBTC system, the Automatic Train Protection (ATP) maintains fail-safe protection against collisions, excessive speed, and other hazardous conditions through a combination of train detection, train separation and interlocking. Computer interlocking (CI) is the interface between CBTC and external trackside field equipment, which establishes the interlocking relationship between signals, switches and routes, in order to avoid the conflicting train operation routes to be set, and to prevent one resource from being occupied by two trains at the same time, both the ATP and CI are safety critical systems which usually meet the highest safety requirements (SIL4). The Automatic Train Operation (ATO) performs the functions of speed regulation, programmed stopping, door control, performance level regulation and other functions otherwise assigned to the train operator. Automatic Train Supervision (ATS) is the subsystem that monitors trains, adjusts the performance of individual trains to maintain schedules, and provides data to adjust service to minimize inconveniences otherwise caused by irregularities. Both the ATO and ATS are safety relevant system which usually have the moderate safety requirements (SIL2).

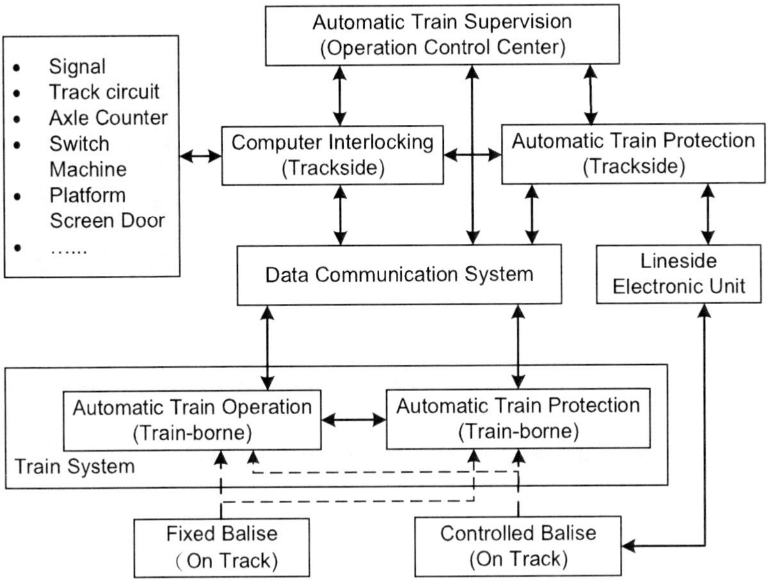

Fig. 1. CBTC system architecture and external interface

In addition to the ATP, CI, ATO and ATS systems, CBTC also includes Data Communication System (DCS) and Balise Transmission System. The DCS is to transmit bi-direction information transmission between trackside equipment and vehicle-mounted equipment. The Balise transmission system is divided into fixed Balise and controlled Balise. The fixed Balise is mainly used for train positioning and position calibration, and controlled Balise is used to transmit some variable temporary commands, such as speed limit information and route information. The external interfaces of the CBTC system include train, platform screen doors, platform emergency buttons, integrated supervision and control system and other systems. These external systems provide input to the CBTC system and execute the control commands issued by the CBTC.

2.2 CBTC System Safety Requirements

As shown in Fig. 2, the ATC system requirements can be divided into safety requirements and non-safety requirements. The safety requirements are derived from the hazard, which are measured via Tolerable Hazard Rate (THR) as the system top level safety goal. The functional safety requirement is part of the safety requirement, it is measured via Tolerable Functional Failure Rate (TFFR), the THR is derived from the TFFR. The functional safety requirements include safety functions and safety integrity, which use the Safety Integrity level (SIL) to define their requirement. The relation between the SIL and TFFR is shown in Table 1. The SIL is the safety performance index used for the defined safety function. The SIL 4 has the highest level of safety integrity. For safety integrity, it includes the Systematic Failure Integrity and Random Failure Integrity and depends on the system design. The systematic failure shall meet the SIL requirement, and the Random Failure shall meet both the SIL and Failure Rate (FR) requirements.

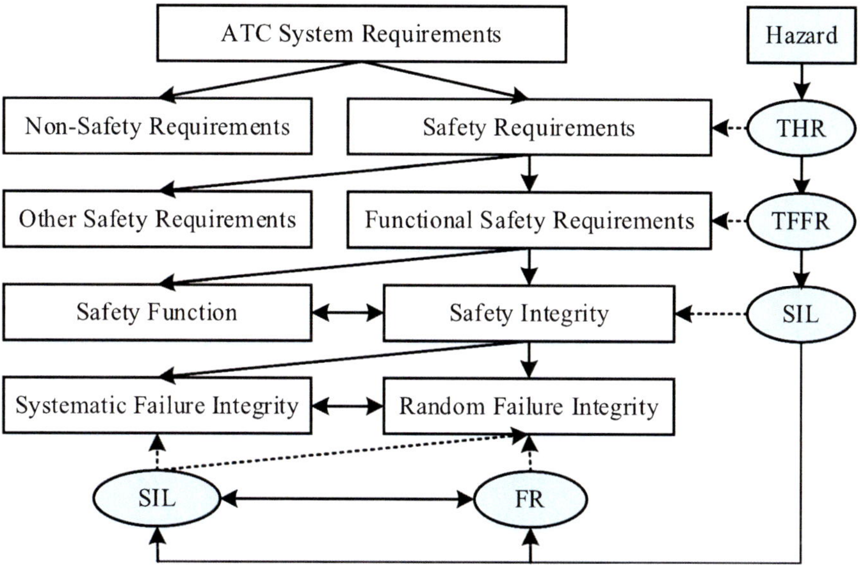

Fig. 2. System Requirements architecture and THR allocation process

In order to ensure the safe train operation, the ATC system shall implement the necessary safety function to prevent the hazard "exceeding speed and/or distance limits advised to ATC, and the THR to this hazard is 10^{-9}/h, the relevant safety function can be decomposed to (1) ensure safe route, (2) ensure safe separation of trains, (3) ensure safe speed, (4) control acceleration and braking [2]. These functions shall meet the TFFR which is allocated from the THR, and if the equipment used to implement these safety function, the equipment should meet the relevant SIL and FR.

As a typical urban rail ATC system, it has following core safety functions: (1)Train location/train speed determination, (2) Safe train separation, (3) Overspeed protection and brake assurance, (4) Rollback protection, (5) End-of-track protection, (6) Parted consist protection and coupling and uncoupling of trains, (7) Zero speed detection, (8) Door opening control protection interlocks, (9) Departure interlocks, (10) Emergency braking, (11) Route interlocking, (12) Traffic direction reversal interlocks, (13) Work zone protection, (14) Broken rail detection, (15) Restricted route protections, (16) Level-crossing protection. These safety function should be implemented by trackside equipment Computer Interlocking (CI) and Zone Controller (ZC) and train borne equipment Onboard Controller (OC), that means the CI, ZC and OC shall meet the SIL 4 requirement with the THR less than THR is 10^{-9}/h [3].

Table 1. The correlation between SIL and TFFR

No.	TFFR per hour and per function	Safety Integrity Level
1	$10^{-9} \leq$ TFFR $< 10^{-8}$	4
2	$10^{-8} \leq$ TFFR $< 10^{-7}$	3
3	$10^{-7} \leq$ TFFR $< 10^{-6}$	2
4	$10^{-6} \leq$ TFFR $< 10^{-5}$	1

3 Safety Requirement Allocation Method

From the process of ATC system implement the safety control function, it shows that if any of the CI, ZC and OC failed, an accident may happen in the urban rail line, which will lead to the overall THR goal cannot be met. From the safety point of view, the CI, ZC and OC of the ATC system form a series system as described in Fig. 3.

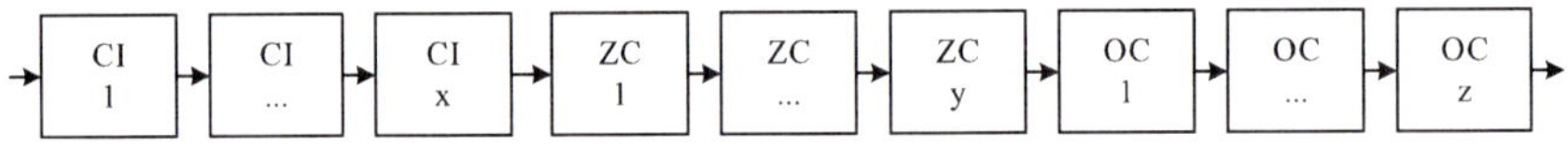

Fig. 3. ATC system safety control logic model

As rail authority may only focus on the safety requirements of ATC system level but not the subsystem CI, ZC, and OC, but for subsystem manufacturers, the safety requirements of ATC system level need to be allocated to CI, ZC and OC subsystems. From the definition of safety requirements, the safety integrity level is divided into two parts: the non-quantifiable SIL and the quantifiable TFFR (FR). For various components that performs one safety function, if there is no redundancy, the components will directly inherit the non-quantifiable SIL from the system, that is, the SIL for CI, ZC, and OC subsystems are the same as the SIL of ATC system, and they are all SIL4 components. For the quantifiable part of the safety requirements, we need have allocation methods to allocate the ATC system THR and TFFR to CI, ZC, and OC and make sure the final total subsystem TFFR is less than the ATC system TFFR and the hazard THR, which is described in Eq. (1).

$$\sum_{i}^{X} TFFR_{CIi} + \sum_{i}^{Y} TFFR_{ZCi} + \sum_{i}^{Z} TFFR_{OCi} \leq TFFR_{ATC} \leq THR \tag{1}$$

3.1 Equal Apportionment Technique

If we ignore the specific subsystem (CI, ZC and OC) detail property and all the subsystems are operated in series as shown in Fig. 3, equal apportionment to each subsystem would seem reasonable. The equal apportionment technique assumes a series of "n" subsystems, each of which is to be assigned the same safety goal [4]. A prime weakness of the method is that the subsystem goals are not assigned in accordance with the degree of difficulty associated with achievement of these goals. For this technique, the model is:

$$S_{ATC} = S_{CI} \times S_{ZC} \times S_{OC} \tag{2}$$

$$TFFR_{CI} = TFFR_{ZC} = TFFR_{OC} = \frac{TFFR_{ATC}}{x + y + z} \tag{3}$$

S_{ATC} means the safety goal of ATC system, and S_{CI}, S_{ZC}, S_{OC} represent the safety goal of CI, ZC and OC subsystems respectively. $TFFR_{ATC}$, $TFFR_{CI}$, $TFFR_{ZC} TFFR_{OC}$ are the TFFR value for ATC, CI, ZC and OC subsystems. The x, y and z are the numbers of CI, ZC and OC deployed in the specific urban rail line separately.

3.2 Safety Impact-Based Apportionment Technique

The ATC system is made up from various subsystems, and the safety impact of the different subsystems to the ATC system is different, that is, if a subsystem has a dangerous failure, it will have different safety impact on ATC system in comparison with others, and it will cause different accident severity. Therefore, different safety goals need to be assigned to different subsystems depending on the safety impact. For subsystems with a bigger safety impact, the safety goal should be more stringent.

For the ATC system, according to the magnitude of the safety impact of the CI, ZC and OC subsystems on the rail line ATC system, the safety impact weight factor is defined in the Table 2. The safety impact number a, b and c mean 1 failure in CI, ZC and OC will cause a, b and c others subsystem enter into dangerous status. Taking the CI subsystem as an example, the safety impact weight factor for the CI can be defined as the ratio of:

$$w_{CI} = \frac{a - (a + b + c)/3}{(a + b + c)/3} = \frac{2a - b - c}{a + b + c} \tag{4}$$

For the ATC system, its safety goal can be expressed as:

$$TFFR_{ATC} = TFFR_{CI} \times x + TFFR_{ZC} \times y + TFFR_{OC} \times z \tag{5}$$

Let $TFFR_a$ as the base TFFR for the subsystem of ATC, then the

$$TFFR_{ATC} = TFFR_a \times (1 + w_{CI}) \times x + TFFR_a \times (1 + w_{ZC}) \times y + TFFR_a \times (1 + w_{OC}) \times z \tag{6}$$

The $TFFR_a$ can be derived from formula (6)

$$TFFR_a = \frac{TFFR_{ATC}}{x + y + z + w_{CI} \times x + w_{ZC} \times y + w_{OC} \times z} \tag{7}$$

Then the allocated TFFR for subsystem are:

$$TFFR_{CI} = TFFR_a(1 + w_{CI}) \tag{8}$$

$$TFFR_{ZC} = TFFR_a(1 + w_{ZC}) \tag{9}$$

$$TFFR_{OC} = TFFR_a(1 + w_{OC}) \tag{10}$$

3.3 Complexity-Based Apportionment Technique

It is true that if a system contains more basic components, its reliability should be worse, and the probability of failure will be higher, so its safety performance will be worse. Therefore, when allocating the THR, we must consider the number of basic components including sub-systems. The more components contained in the subsystem, the less safety goal should be assigned to that subsystem.

Assuming that the subsystems in ATC is composed of same basic components, these basic components have the same failure rate, so we can use the number of basic components contained in a subsystem to express the complexity of the subsystem. We use the numbers l, m and n to represent the complexity of CI, ZC and OC subsystems. For a specific urban rail ATC system containing x CIs, y ZCs and z OCs, the complexity factor can be expressed as Eqs. (11), (12) and (13).

$$cw_{CI} = \frac{x \times l}{x \times l + y \times m + z \times n} \tag{11}$$

$$cw_{ZC} = \frac{y \times m}{x \times l + y \times m + z \times n} \tag{12}$$

$$cw_{OC} = \frac{z \times n}{x \times l + y \times m + z \times n} \tag{13}$$

Therefore, the assigned TFFR value of a single CI, ZC and OC are described as follows:

$$TFFR_{CI} = \frac{TFFR_{ATC} \times cw_{CI}}{x} = \frac{TFFR_{ATC}}{x \times l + y \times m + z \times n} \times l \tag{14}$$

$$TFFR_{ZC} = \frac{TFFR_{ATC} \times cw_{ZC}}{y} = \frac{TFFR_{ATC}}{x \times l + y \times m + z \times n} \times m \tag{15}$$

$$TFFR_{CI} = \frac{TFFR_{ATC} \times cw_{OC}}{z} = \frac{TFFR_{ATC}}{x \times l + y \times m + z \times n} \times n \tag{16}$$

4 Apportionment Result Comparation

In order to verify the difference and applicability for the allocation results of the THR using the three different allocation methods, this research applies the allocation method proposed in the previous section to calculate the allocation results, as well as the ATC system THR allocation results are analyzed in a specific urban rail project.

4.1 Assumptions of the Specific ATC System

As shown in Fig. 4, in the specific ATC system of urban rail transit project, only the CI, ZC and OC are responsible for the vital safety functions. The ATC system includes the number of CI, ZC and OC subsystems which are x, y and z, where x is equal to 6, y is equal to 3, z is equal to 30. For the entire ATC system, the overall safety goal proposed by the operator is that THR should be less than 10^{-9}/h.

For the safety impact of the CI, ZC, and OC subsystems, we define one failure in CI, ZC, and OC subsystems that can impact the number of other subsystems are a, b and c. where a = 4, b = 10, and c = 2.

For the complexity of the system, we define each CI, ZC and OC subsystem is composed by l, m and n basic components respectively, where l is equal to 10, m is equal to 8, and n is equal to 12.

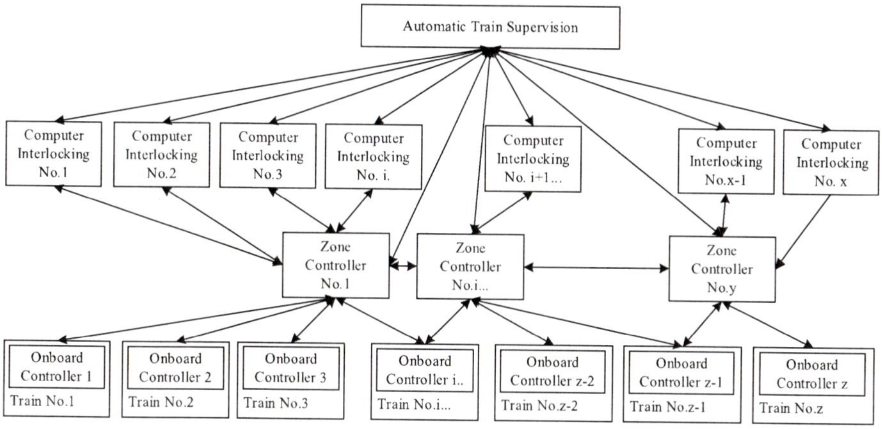

Fig. 4. A Specific ATC System Configuration

4.2 Safety Requirement Allocation Result

Based on the above assumptions the specific ATC System, we used the three different methods to allocate the THR as the ATC system safety requirements in Sect. 3. The ATC system data of calculation required is show in Table 2, and the allocation results are shown in Table 3.

From the above allocation results shown in Table 3, it can be seen that the allocation results of the dangerous failure rate (FR) of each subsystem obtained by using three different methods are all in the same order of magnitude. However, the specific values have certain differences, it can show that all the three methods are feasible. Comparing the calculation results with the actual situation of urban rail project, we think that the allocation method based on complexity is most reasonable. The THR Allocation Results of CI, ZC, and OC subsystems for Complexity-based Apportionment method are less than the EqualApportionment method as well as Safety Impact-based Apportionment method. For the method based on safety impact, since both the overall number

Table 2. The ATC system data

Subsystem	Subsys No.		Safety Impact		Complexity	
Computer Interlocking (CI)	x	7	a	4	l	10
Zone Controller (ZC)	y	3	b	10	m	6
Onboard Controller (OC)	z	30	c	2	n	12

Table 3. THR Allocation Results

Subsystem	Equal	Safety Impact-based	Complexity-based
Computer Interlocking (CI)	2.5×10^{-11}	3.39×10^{-11}	2.23×10^{-11}
Zone Controller (ZC)	2.5×10^{-11}	8.47×10^{-11}	1.34×10^{-11}
Onboard Controller (OC)	2.5×10^{-11}	1.69×10^{-11}	2.68×10^{-11}

of subsystems and safety impact are considered in the allocation, the final allocation result is more sensitive to the number of subsystems than the safety impact. Except the Complexity-based Apportionment method, the other two method lead the bigger safety impact had been assigned a relatively loose safety value. Therefore, we consider the allocation method based on system complexity is the most suitable for the actual conditions of the project, which not only can meet the requirements of the Urban Rail Automatic Train Control System, but also can improve the reliability and safety of urban rail transit to some extent.

5 Conclusions

From the above analysis, it can be concluded that using the ATC system level THR as the safety requirements of the subsystems without any allocation cannot meet the overall system safety goals of the rail transit line. Therefore, using the different THR allocation methods will result in different subsystem TFFR, but the same order of magnitude. By comparing with the actual engineering demands, the allocation method based on complexity is the more applicable than the other two allocation methods, due to the allocation method based on complexity considers the specific characteristics of the subsystems.

Eventually, no matter which allocation method is adopted, we need to consider not only the logical architecture of the system, but also the specific number of each subsystem in the line system. At last, a reasonable allocation result of Complexity-based Apportionment Technique can be obtained. It is indicated from the results of the allocation that the larger the scale of the rail transit line is, the more subsystems it contains, and the more stringent and stricter safety requirements for the subsystems should be allocated.

Acknowledgments. The project is supported by Science and Technology Commission of Shanghai Municipality (20DZ2251900), as well as the Natural Science Foundation of Shanghai

(21ZR1423800). The work is also sponsored and supported by the Key Laboratory of Road and Traffic Engineering, Tongji University and Shanghai Engineering Research Center of Urban Infrastructure Renewal, Shanghai University. The authors are grateful for the reviewer of initial drafts for their helpful comments and suggestions.

References

1. IEEE, IEEE 1474.1, IEEE Standard for Communication-Based Train Control(CBTC) performance and functional Requirements. IEEE, New York (2004)
2. IEEE, IEEE 1474.3, IEEE Recommended Practice for Communication-Based Train Control(CBTC) System Design and Functional Allocations. IEEE, New York (2008)
3. CENELEC. EN 50129, Railway applications: safety related electronic systems for signaling. 3.CENELEC, Brussels, (2018)
4. Department of, Defense: MIL-HDBK-338B. Electronic Reliability Design Handbook. Department of Defense, New York (1998)

Research on the Linkage Method of Each Link of Urban Public Transport Service Purchased by Government

Jiang Xiantong[1] and Wang Chang[2](✉)

[1] China Academy of Transportation Science, Chaoyang, Beijing 100029, China
[2] Transportation Science Technology Consulting (Beijing) Co. Ltd., Chaoyang, Beijing 100029, China
595402427@qq.com

Abstract. At present, China's cities still have the situation that all links of the government's purchase of urban public transport services are relatively separated, and the efficiency of public transport services and the use of financial funds is not high. Therefore, this paper proposes a linkage method based on all links of public transport service purchased by the government, including public transport operation service quality standard, cost standard, fare level, service assessment method, and comprehensive performance and so on. By taking the completion of performance objectives such as quantity, quality, cost and benefit of public transport enterprises as a reasonable basis for the actual payment of service fees purchased by the government, guide public transport enterprises to actively improve quality and efficiency and achieve the highest service level. Finally, the study selects a city to apply the method, and issues a landing policy document. The application results show the effectiveness of the method and improve the public transport service level and the use efficiency of financial subsidy funds in the city.

Keywords: Urban public transportation · Government's purchasing healthcare service · Linkage · Financial subsidies

1 Introduction

According to the measures for the administration of government procurement of services, state organs at all levels are responsible for the services that belong to their own responsibilities and are suitable for market-oriented provision to qualified units and enterprises, and they must pay fees according to the methods and procedures of government procurement and the factors such as the quantity and quality of services [1]. As a social public welfare undertaking to meet the basic travel needs of the people, urban public transport belongs to the scope of public services purchased by the government.

In fact, with the rise of labor, price and other costs and the continuous implementation of low ticket price policy in most cities in China, the vast majority of public transport enterprises in China are unable to cover the cost only by ticket income, and the government needs to make up for the loss of public transport operation through financial

X. Zeng et al. (Eds.): ITASC 2022, LNEE 1042, pp. 136–146, 2023.
https://doi.org/10.1007/978-981-99-2252-9_13

subsidies. The traditional public transport financial subsidies in China include: one case one discussion subsidy, bargaining subsidy, special subsidy, and bottom-up subsidy, but there are some common problems: firstly, the service standard is decided by the industry management department or the public transport enterprises themselves, which is not comprehensive in considering the public transport financial affordability and meeting the basic public travel services; secondly, the public transport cost standard is decided by the public transport enterprises themselves, which is basically legal and compliant, but not necessarily reasonable under the constraints of the annual audit; thirdly, the relationship between the public transport financial subsidies and the service level is not strong, so enterprises are not willing to take the initiative to improve the quality and efficiency; fourth, the relationship between public transport financial subsidies and the ticket revenue determined by public transport passenger volume and ticket price is not clear, to a certain extent, the pressure of passenger volume increase or decrease due to various reasons is on the public transport enterprises.

In the implementation of traditional public transport financial subsidy policy, public transport enterprises think that the financial subsidies are not in place, which leads to serious losses of public transport enterprises; however, the government financial departments think that the public transport financial subsidies are too much and the capital pressure is too high. With the increasing gap of public transport operation funds, the contradiction between the government and public transport enterprises is becoming more and more prominent. Therefore, it is necessary to reconcile the contradiction between the government and enterprises through certain methods. In view of the current situation that all links are not closely linked, we can establish rules through the concept of government purchasing services, and form the linkage method of public transport service standards, costs, assessment, ticket prices and subsidies. First, we should clarify the responsibilities of industry management departments and public transport enterprises, consider the public transport financial affordability and operation efficiency, meet the public travel needs, and formulate the public transport operation plan Service standards; second, through cost regulation to require the legal and compliance of public transport operation costs, and formulate public transport operation service standards; third, link the customization and adjustment of public transport fares with the cost level; fourth, link the performance of public transport assessment with the completion of the quantity and quality standards of public transport services; fifth, the performance of public transport assessment directly affects the actual subsidy of public transport finance, which will effectively supervise the public transport Enterprises should take the initiative to improve quality and efficiency.

2 Technical Route

Through the establishment of public transport operation service standards, the relationship between service assessment results and performance is constructed, the revenue of public transport tickets linked with passenger ticket price and the costs are calculated, and the public transport financial subsidies are finally determined. The policy formulation and implementation of each stage will be fed back to the new round of service purchase policy research and formulation, as shown in the Fig. 1 below.

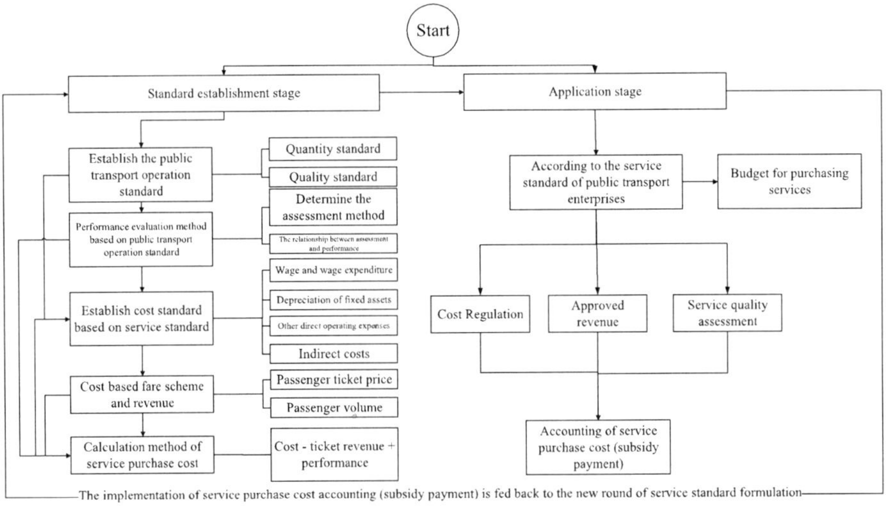

Fig. 1. Flow chart of subsidy linkage method of public transport service standard, cost, assessment and ticket price based on government purchase service

3 Establish Bus Operation Standards

3.1 Establish the Service Standard and Assessment Standard of Public Transport Operation

Based on the data of urban public transport enterprises over the years and the forecast of passenger flow demand, combined with the government's financial budget, this paper sets the total requirements and quantity standards of urban public transport enterprises, and reserves some emergency budget.

According to the characteristics of the city, the quantitative and quality indicators of public transport operation service assessment are established, and the corresponding assessment standards are set.

3.2 Establishing the Regulation Method of Public Transport Operation Cost

The cost of establishing the regulation method of public transport operation cost is closely related to the standard of public transport operation service. On the one hand, the quality standard of public transport service reflects the unit cost price of each cost item; On the other hand, the quantity standard of public transport service reflects the supply of public transport operation service. Therefore, the calculation formula of bus operation cost is as follows:

$$\mathrm{Cg} = \mathrm{Rg} + \mathrm{Gg} + \mathrm{Zg} + \mathrm{Jg} \tag{1}$$

where: Rg refers to the annual salary and wage expenditure, which is related to the standard of the number of public transport operations provided. The formula is as follows:

$$\mathrm{Rg} = \left[\frac{\mathrm{S}}{\mathrm{Tg}} \times \mathrm{W} \times (1 + \mathrm{a}) + \frac{\mathrm{S}}{\mathrm{Tg}} \times \mathrm{vg} \times \mathrm{W} \times (1 + \mathrm{b})\right] \times (1 + \mathrm{u}) \tag{2}$$

where: s is the total working time of the driver, including passenger carrying working time and empty driving working time, Tg is the standard working time of the driver, W is the reference wage standard of the driver, Vg is the ratio of the management assistant and the driver, a is the wage coefficient of the driver, b is the wage coefficient of the management assistant, u is the proportion of five insurances and one fund, employee welfare, labor union funds, employee education funds, etc. These parameters are set according to the law or the data of local enterprises in the first three years.

Gg is the annual depreciation of fixed assets such as infrastructure and vehicles.

Zg refers to other direct operating expenses, including fuel and power consumption, warranty fee, tire fee, insurance fee, etc. The formula is:

$$\mathrm{Zg} = \mathrm{Bg} \times \mathrm{L} \tag{3}$$

where: Bg is the cost standard per kilometer, l is the standard of bus operation quantity, i.e. operation mileage.

Jg is the annual indirect operating cost, including administrative expenses.

In summary, the calculation formula of bus operation cost based on the bus service standard is:

$$\begin{aligned}\mathrm{Cg}=&\left[\frac{\mathrm{S}}{\mathrm{Tg}} \times \mathrm{W} \times (1+\mathrm{a}) + \frac{\mathrm{S}}{\mathrm{Tg}} \times \mathrm{v} \times \mathrm{W} \times (1+\mathrm{b})\right] \times (1+\mathrm{s})\\ &+ \mathrm{Gg} + \mathrm{Bg} \times \mathrm{L} + \mathrm{Jg}\end{aligned} \tag{4}$$

4 Research on the Linkage Method of Purchasing Public Transport Services

4.1 Linkage Between Bus Fare Revenue and Passenger Fare

Public transport ticket revenue is the main revenue of public transport operation services. The level of ticket revenue directly affects the amount of financial subsidies. The relationship between passenger ticket price and public transport ticket revenue is as follows:

$$\mathrm{Ps} = \mathrm{Pr} \times \mathrm{N} \tag{5}$$

where: Ps is the ticket revenue, Pr is the ticket price per person per time, which is affected by the proportion of ticket price and cost, the proportion of financial subsidies in general public budget expenditure, and the proportion of residents' commuting expenses in disposable income; n is the passenger volume.

4.2 Linkage of Service Assessment Results and Performance

According to the data and service quality assessment of local enterprises in the recent three years, combined with the set bus operation service quantity standard and quality standard, the relationship between quantity assessment results, quality assessment results and performance is established. The formula is as follows:

$$\mathrm{Jk} = \mathrm{Jks} + \mathrm{Jkz} \tag{6}$$

where: Jk is the evaluation performance; Jks is the quantity evaluation performance; jkz is the quality evaluation performance.

The quantity evaluation performance is inversely linked with the evaluation results: no reward or punishment for completing the task; one penalty for failing to meet the task requirements To impose a fine, that is:

$$\text{When } \mathrm{Sc} - \mathrm{Sct} \geq 0, \mathrm{Jks} = 0 \tag{7}$$

$$\text{When } \mathrm{Sc} - \mathrm{Sct} < 0, \mathrm{Jks} = (\mathrm{Sc} - \mathrm{Sct}) \times (\mathrm{Cg} - \mathrm{Ps}) \times \mathrm{c} \tag{8}$$

where: Jks is the quantitative assessment performance; Sc is the quantitative assessment result; Sct is the qualification standard of quantitative assessment result; C is the proportion of quantitative performance linked with regulation cost and actual income difference.

The quality evaluation performance and evaluation results are positively and negatively linked: excellent task completion, there is a certain reward; qualified task, no reward and no punishment; unqualified task, there is a certain fine, that is:

$$\text{When } \mathrm{Zc} > \mathrm{Zct1}, \mathrm{Jks} = (\mathrm{Zc} - \mathrm{Zct1}) \times (\mathrm{Cg} - \mathrm{Ps}) \times \mathrm{d} \tag{9}$$

$$\mathrm{Zct2} \leq \mathrm{Zc} \leq \mathrm{Zct1}, \mathrm{Jks} = 0 \tag{10}$$

$$\text{When } \mathrm{Zc} < \mathrm{Zct2}, \mathrm{Jks} = (\mathrm{Zc} - \mathrm{Zct2}) \times (\mathrm{Cg} - \mathrm{Ps}) \times \mathrm{d} \tag{11}$$

where: Jks is the performance of quality assessment; Zc1 is the result of quality assessment; Zc is the upper limit of qualified standard of quality assessment; Zc2 is the lower limit of qualified standard of quality assessment; d is the proportion of quality performance linked with regulation cost and actual income difference.

4.3 Linkage of Public Transport Financial Subsidies with Performance, Cost and Ticket Price

Establish a cost regulation method based on operating service standards, a method for calculating bus operating revenue based on passenger fares and passenger volume, and a method for determining subsidies based on service assessment to calculate public transport financial subsidies. The calculation formula is:

$$\mathrm{Bg} = \mathrm{Cg} - \mathrm{Ps} + \mathrm{Jk} \tag{12}$$

where: Bg is the bus operation subsidy based on the bus service standard, Cg is the bus operation regulation cost based on the bus service standard, Ps is the actual ticket income, and Jk is the performance appraisal.

5 Method Application Case

This method is applied in a city with two bus operators. By setting the bus service standard, the budget of bus operation cost is set based on the bus service standard, and then the performance and bus operation income are obtained through the assessment, and the actual bus operation subsidy is calculated.

5.1 Bus Operation Cost Parameter Setting

According to the requirements of the labor law, referring to the social average wage and price level of the regulation year, and combined with the requirements of the total amount of public transport services, the parameters of public transport operation cost are set:

$$Cg=\left[\frac{S}{1920\ \text{hours}} \times W \times (1 + 120\%) + \frac{S}{1920\ \text{hours}} \times v(0.5) \times W \times (1 + 108\%)\right] \times (1 + u) + m \times L + Gg + Jg \quad (13)$$

where, the meaning of the letter is the same as above.

5.2 The Linkage Between the Evaluation of Public Transport Operation Service and Performance

According to the data and service quality assessment of local enterprises in the recent three years, the indicators and standards of public transport service quantity assessment and quality assessment are set, so as to calculate the performance of public transport operation service quality assessment.

(1) Establish the Standard of Quantity and Quality Assessment

The standards for the number of public transport services are set as follows (Table 1):

Table 1. Evaluation standard of public transport service quantity

Assessment form	Evaluation index	Annual standard	Full marks
Daily assessment	Driving rate in peak hours	Not less than 95%	50
	On time departure rate in peak hours		50
Total score			100
Annual assessment	Completion rate of planned passenger mileage	Not less than 95%	50

(*continued*)

Table 1. (*continued*)

Assessment form	Evaluation index	Annual standard	Full marks
	Driving rate in peak hours		50
Total score			100
Third party assessment	Bus passenger satisfaction	80% and above	100
Total score			100

Quantity assessment result = daily assessment quantity index score × 35% + annual assessment quantity index score × 50% + Third party assessment score × 15%

The standard of public transport service is set, as shown in the Table 2 below:

Table 2. Evaluation standard of public transport service quality

Assessment form	Evaluation index	Annual standard	Full marks
Daily assessment	Completion rate of service facilities	Not less than 90%	20
	Qualified rate of operation service		25
	Pass rate of vehicle neatness		25
	Actual allocation rate of drivers	100%	30
Total score			100
Annual assessment	Standard reaching rate of first and last shift of line operation	Not less than 97%	25
	Bus failure rate	Less than 10 times / million km	17
	Death rate of bus liability accident	Less than 0.02 person / million km	20
	Bus liability accident rate	Less than 3 times / million km	12
	Ratio of complaint cases to operating mileage	Less than 50 pieces / million km	16
	Completion rate of complaint handling	100%	10
Total score			100
Third party assessment	Bus passenger satisfaction	80% and above	100
Total score			100

Quality assessment result = daily assessment quantity index result × 25% + annual assessment quantity index result × 60% + Third party assessment result × 15%

(2) **Establish the Relationship Between Service Assessment Results and Performance**

In the relationship between quantity assessment and performance, the critical line of setting whether to deduct expenses is 95 points, and the linked value is 0.3%. In the relationship between quality assessment and performance, setting more than 95 points is excellent, with a certain reward; 85–95 points is qualified, without award or punishment; 85 points below is unqualified, with a certain fine, with a linked value of 0.3%.

5.3 Calculation of Public Transport Financial Subsidies

(1) **Determine the Completion of Bus Operation**

Combined with the input of public transport network and the level of passenger traffic over the years, the operating mileage, passenger carrying mileage, working time standard and actual completion of public transport are shown in the Table 3 below.

Table 3. Bus operating mileage, passenger carrying mileage, working time standard and actual completion

Mileage	Public transport enterprise 1	Public transport enterprise 2	Total
Planned operating mileage (10000 km)	22052	15618	37670
Actual operating mileage (10000 km)	21170	15261	36431
Completion rate	96.00%	97.71%	96.71%
Planned passenger mileage (10000 km)	21514	14874	36388
Actual passenger mileage (10000 km)	20593	14526	35119
Completion rate	95.72%	97.66%	96.51%
Total planned passenger carrying time (hours)	14641572	9110500	23752072
Actual total passenger carrying time (hours)	14151070	8763129	22914199
Completion rate	96.65%	96.19%	96.47%

(2) Regulating Public Transport Operation Cost

According to the calculation method of personnel wage and wage expenditure cost, and the calculation method of bus operation cost, the comprehensive operation cost of the two enterprises, as shown in Table 4 below.

Table 4. Regulation of comprehensive operation cost of public transport

Cost item	Public transport enterprise 1	Public transport enterprise 2	Total
Personnel wages and wage expenditure (billion yuan)	1.48	0.80	2.29
Depreciation of fixed assets (billion yuan)	0.02	0.01	0.03
Other direct operating costs (billion yuan)	0.55	0.40	0.95
Indirect cost (billion yuan)	0.04	0.02	0.07
Total (billion yuan)	2.09	1.24	3.33

(3) Calculate the Actual Ticket Income

Combined with the proportion of ticket price and cost, the proportion of financial subsidies in general public budget expenditure, and the proportion of residents' commuting expenses in disposable income, the average ticket price per person in the city is set (Table 5).

Table 5. Actual ticket revenue of bus operation

Item	Public transport enterprise 1	Public transport enterprise 2	Total
Average person time ticket price (yuan / person time)	0.8	0.82	0.80
Passenger traffic volume (billion person times)	0.69	0.35	1.03
Bus ticket revenue (billion yuan)	0.55	0.28	0.83

(4) Get Assessment Results

The quantity and quality appraisal results are shown in the Table 6 below.

Table 6. Summary of bus assessment results

Achievement	Public transport enterprise 1	Public transport enterprise 2
Quantity assessment results (points)	94.5	96.9
Quality assessment results (points)	96.8	78

(5) Calculation of Actual Public Transport Financial Subsidies

According to the above data, the performance of public transport operation service is calculated, and the subsidies of the two public transport enterprises are shown in the Table 7 below.

Table 7. Two public transport financial subsidies

Item	Public transport enterprise 1	Public transport enterprise 2	Total
Bus operation cost (billion yuan)	2.09	1.24	3.33
Bus ticket revenue (billion yuan)	0.55	0.28	0.83
Cost revenue (billion yuan)	1.54	0.95	2.49
Quantity performance (billion yuan)	−0.002	0.00	−0.002
Quality performance (billion yuan)	0.008	−0.020	−0.008
Actual financial subsidy (billion yuan)	1.544	0.933	2.481

6 Conclusion

According to the actual financial subsidies of the city's public transport operation over the years, the amount of public transport financial subsidies estimated by this method in 2019 has a certain increase compared with previous years, but it belongs to a reasonable growth range, and this method establishes the linkage relationship of standard, cost, ticket price and subsidies, which makes the estimated financial subsidies more standardized Scientific and scientific, control the growth rate of bus operating costs and subsidies. The policy document was successfully released and implemented, and the effectiveness of the linkage method based on the government's purchase of public transport services was verified. At the same time, for government departments, the research and application of this method can provide solid data support for government departments to grasp the operating cost of public transport, public transport enterprises to guarantee operating funds and improve quality and efficiency: Through the determination of the standard, the budget rigidity of the financial department for financial subsidies is improved, and the reasonable linkage relationship between operation service and cost subsidies is established, which provides a powerful grasp for the industry management department to improve the level of public transport operation service. When making the annual operation plan, public transport enterprises should not only consider the residents' travel service, but also fully consider the public transport operation cost, fare revenue and subsidy budget, so as to ensure the financial sustainability of public transport operation and the satisfaction of passengers.

References

1. Measures of the Ministry of finance of the people's Republic of China for the administration of government purchase of services (Order No. 102 of the Ministry of finance of the people's Republic of China) [EB / OL] (2020-1-3)
2. Wang, J., Zhong, C., Yu, K.: Research on the design of urban public transport government purchase service system based on the linkage of cost, income and subsidy. Transp. Manager World **3**, 76–79 (2017)
3. Zhong, Z.: Theoretical exploration and promotion path of government purchasing public transport services. Transp. Eng. **6**, 51–57 (2018)
4. Wang, J., Xu, C.: Public transport operation service quality assessment mode design from extensive to refined. Transp. Manager World **4**, 86–91 (2017)
5. He, Q.: Research on government purchase of public services. Beijing: Institute of financial Sciences, Ministry of finance (2014)
6. Wang, P.: Analysis on the reform mechanism of government purchasing public services from social forces. J. Peking Univ. (Philosophy and Social Sciences) **7**, 88–94 (2015)

Research on Regulation Method of Bus Driver's Total Wages Based on Labor Intensity and Operation Efficiency – A Case Study of Yantai City

Xiantong Jiang[1] and Chang Wang[2](✉)

[1] China Academy of Transportation Science, Chaoyang, Beijing 100029, China
[2] Transportation Science Technology Consulting (Beijing) Co., Ltd., Chaoyang, Beijing 100029, China
595402427@qq.com

Abstract. In view of the common inconsistency between salary and labor intensity of bus drivers in China, combined with the requirements of the latest policy documents issued in the fields of domestic public transport and purchase of services, aiming at solving the problem of sustainable development of public transport, improving the level of public transport services and improving the efficiency of capital use, and comprehensively considering the labor intensity and operation efficiency of bus drivers, Optimize the regulation method of bus driver's total salary. This method reflects the idea that the wage expenditure of public transport operation is inclined to front-line drivers, while taking into account fairness and efficiency. Finally, taking Yantai City as an example, this paper calibrates the parameters of the regulation method and writes it into the official policy document of Yantai public transport purchase service implementation plan (2021–2023). This result can be used as a reference for government departments and social institutions in similar cities.

Keywords: Public transportation · cost regulation · bus drivers · wages

1 Introduction

With the establishment and acceptance of two batches of "national transit cities" by the Ministry of transport, the development of public transport in China has entered the "post transit city era", and the national strategy of giving priority to the development of urban public transport has become the consensus to guide the development of cities; public transport has also entered the stage of high-speed development, the infrastructure, capacity and service quality of urban public transport have been improved, and the people's livelihood has been improved The degree of satisfaction and sense of gain of the students were greatly improved. However, the bottleneck of sustainable development of public transport has not been solved: (1) the policy deficit of public transport operation is increasing rapidly, the local financial pressure is increasing year by year, and the

X. Zeng et al. (Eds.): ITASC 2022, LNEE 1042, pp. 147–156, 2023.
https://doi.org/10.1007/978-981-99-2252-9_14

total amount of financial subsidies is growing out of control; (2) public transport enterprises are generally in a state of loss, and the endogenous power of saving expenditure, improving quality and efficiency is insufficient.

In order to effectively solve the above problems and realize the sustainable development of public transport, cities begin to resort out and clarify the responsibilities and rights of all participants in public transport, and actively explore the purchase service mode, capital scheme, operation system and mechanism; among them, the reasonable regulation of public transport operation cost is the main basis for the implementation of public transport purchase service, which is to alleviate the burden of capital. It is an important way to realize the contradiction between the financial subsidy and the development of public transport, and to reflect the rationality, legality and fairness of financial subsidies. For most cities, the salary and wage expenditure of public transport practitioners account for more than 50% of the total operating cost, which has been the focus of attention of all parties. This paper focuses on the regulation method of total salary of public transport main practitioners - bus drivers, and takes Yantai City as an example, based on the labor intensity and operation efficiency of local bus drivers, puts forward some suggestions.

2 Technical Thinking

The specific technical ideas are shown in the figure below (Fig. 1):

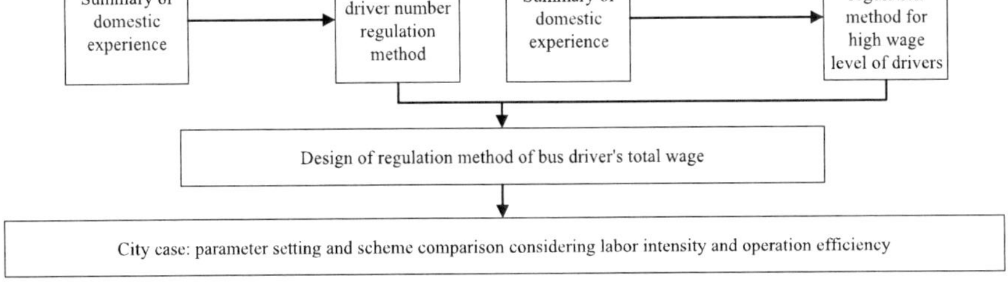

Fig. 1. Research technology route

3 Regulation Method of the Number of Drivers

3.1 The Domestic Common Regulation Method of the Number of Drivers

At present, there are three main methods to regulate the number of drivers in China:

Method 1: Regulate the Number of Drivers Through the Average or Weighted Average of the Actual Ratio of People to Vehicles in Recent Three Years Shijiazhuang, Changzhou and other cities adopt the single index "comprehensive ratio of people to vehicles", while Tianjin adopts "ratio of drivers to vehicles". The formula is: the number of regulated drivers = the number of vehicles × the proportion of regulated drivers. Due to the technological progress and the improvement of enterprise management level, the ratio of people to vehicles is declining year by year. This method has certain rationality and is easy to operate. However, the existing problems are insufficient constraints, different working intensity of drivers in different cities, different road operation environment, different turnover efficiency of drivers and vehicles, and the number of vehicles can not reflect the labor intensity and transportation efficiency Business efficiency is not comparable among different cities.

Method 2: Regulate the Number of Drivers According to the Vehicle Operation Time The required drivers are regulated by departure frequency, legal working days and average number of vehicles. The calculation formula is: average daily departure times of that year × 365/legal working days/annual average number of vehicles. Compared with the first calculation method, this calculation method has the following advantages: it reflects the efficiency of enterprise scheduling and vehicle utilization, and solves the driver vehicle ratio problem. The premise of using this method to calculate the driver vehicle ratio is that the enterprise scheduling and transport capacity allocation are relatively reasonable and efficient.

Method 3: Regulate the Number of Drivers According to the Working Hours of the Whole Year By calculating thc ratio of thc total annual working hours of drivers to the standard working hours of drivers, the number of drivers is regulated. The calculation formula is: total working hours of drivers on all operating lines/annual working hours of drivers per capita. The standard of annual per capita working hours is 1880 h/person · year - 2432 h/person · year, which is also the calculation method recommended in the industry standard "urban bus and tram enterprise operating cost calculation specification". The advantage of this method is that it fully considers the labor intensity of the driver, and combines with the labor law and other standards to encourage bus drivers not to work overtime. However, there are some risks in this method: (1) regarding the driver's working time as the total operation time, the corresponding efficiency constraints are insufficient; (2) there is a risk that the driver deliberately reduces the operation efficiency, leading to the cost increase.

3.2 Design of Regulation Method to Guide the Number of Drivers to Improve Efficiency

With the development of public transportation informatization, it's easy to obtain the operating mileage and working hours of drivers. Therefore, we can make full use of the relationship between the total working hours and the standard working hours in the industry standards to overcome the problems of drivers deliberately reducing the speed and increasing the non-passenger time, and establish the regulation formula of the number of drivers as follows:

$$JR = \frac{\frac{ZL}{S}/(1-\mathrm{f})}{Rt} \tag{1}$$

where:

JR——Number of drivers regulated;

ZL——Annual passenger working mileage;

S——Annual average passenger carrying speed, which is related to the degree of bus lane and road congestion in each city, and it is recommended to calculate according to the historical average level;

f——The ratio of non-passenger working time to total operation time is suggested to be 20%–30%;

Rt——The standard working hours of the whole year refer to the industry standard, and the value range is 1880 h/person–2432 h/person.

4 Regulation Method of Driver's Wage Level

4.1 The Domestic Common Regulation Method of Driver Wage Level

For the regulation method of driver level, the domestic cities are basically the same, and the general calculation formula is as follows:

$$\mathrm{JZ} = \mathrm{W} \times (1 + \mathrm{Rg}) \times (1 + \mathrm{a}) \tag{2}$$

where:

JZ——Driver's salary level;

W——the reference salary benchmark, which is generally the average wage of the previous year issued by the government in the regulatory year;

Rg——With reference to the benchmark wage growth rate, which is generally taken according to the average growth rate of the wage benchmark in recent three years;

a——Driver's wage factor.

4.2 Design of Driver Wage Regulation Method Considering Growth Regulation Mechanism

According to the "opinions of the State Council on reforming the wage determination mechanism of state owned enterprises" issued in 2018, the regulation method of driver's wage level linked with public transport service quality assessment is designed.

$$\mathrm{JZ} = \mathrm{W} \times (1 + \mathrm{Rg} \times \mathrm{b}) \times (1 + \mathrm{a}) \tag{3}$$

where, the same index is consistent with the above. Besides, b——The adjustment coefficient of driver's salary. When the examination result is excellent, the value is 1–1.5; when the examination result is qualified, the value is 0.5–1; when the examination result is unqualified, the value is 0. The assessment of public transport should include cost control, service quality, operation efficiency, support ability and other indicators.

This can effectively avoid the problem that the wage growth is divorced from the service quality assessment, effectively promote the bus companies to continuously improve the level of cost control and service quality, and effectively improve the efficiency of the use of financial funds.

5 Regulation Method of Bus Driver's Total Wage

Based on the results of the regulation of the number of drivers and the regulation of the wage level, the calculation formula of the total wage of drivers is as follows:

$$JZ = JR \times JG = \left[\frac{\frac{ZL}{S}}{\frac{1-f}{RT}} \right] \times [\mathrm{W} \times (1 + \mathrm{Rg} \times \mathrm{b}) \times (1 + \mathrm{a})] \tag{4}$$

where, the same index is consistent with the above.

6 Formula Parameter Value of Comprehensive Labor Intensity and Operation Efficiency - Taking Yantai City as an Example

6.1 Status Assessment

6.1.1 Labor Intensity of Bus Drivers

According to Yantai's bus intelligent system, the annual working hours of drivers are 3268 h, far higher than the recommended value of 1880 h/person–2432 h/person in the industry standard. In addition, through horizontal comparison of the average operating mileage of drivers in various cities, Yantai reached 54817 km/person in 2019, which is much higher than that in other cities. As shown in the figure below (Figs. 2, 3 and 4).

Fig. 2. Comparison of per capita operating mileage of drivers in different cities

6.1.2 Salary Level of Bus Drivers

In 2016–2019, the salary level of bus drivers in Yantai city is 102–112% of the average wage level of Urban Non private enterprises, which is at a high level; however, the average wage converted into 1920 h (the recommended value of cost regulation industry standard) is only 59–65% of the average wage level of Urban Non private enterprises. In contrast, the salary level of bus drivers in Qingdao, Jinan and other cities can reach about 90%. Therefore, under the same labor intensity, the salary of bus drivers in Yantai is at a lower level.

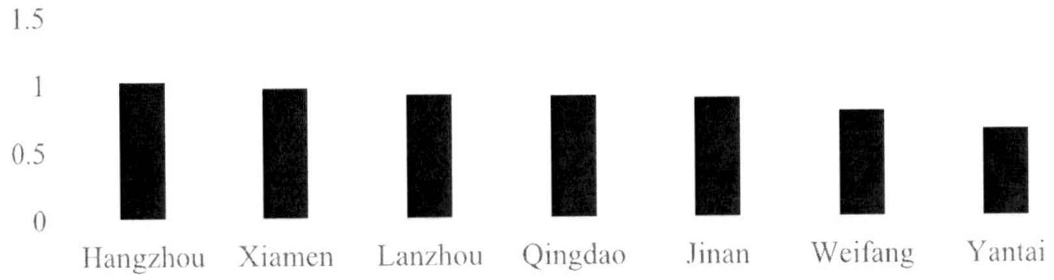

Fig. 3. Converted into 1920 h driver wage/average wage of non-private enterprises in local cities and towns in the current year

6.1.3 Bus Driver Efficiency

The work efficiency of bus drivers mainly involves the setting of two parameters: annual average passenger carrying speed and the ratio of non-passenger working time to total working time.

1. The average passenger carrying speed of Yantai is 23 km/h, which is in the medium level and can be taken as a parameter;
2. Yantai's non passenger working hours (refueling, gas filling, in and out, waiting for duty, etc.) account for 31.4% of the total operation time, which is higher than other cities. This is due to the relatively low efficiency of scheduling. Drivers have long operating mileage and long non passenger carrying time, so the labor intensity is high. Therefore, in the selection of regulatory parameters, we should guide to reduce the labor intensity of drivers and improve the efficiency at the same time.

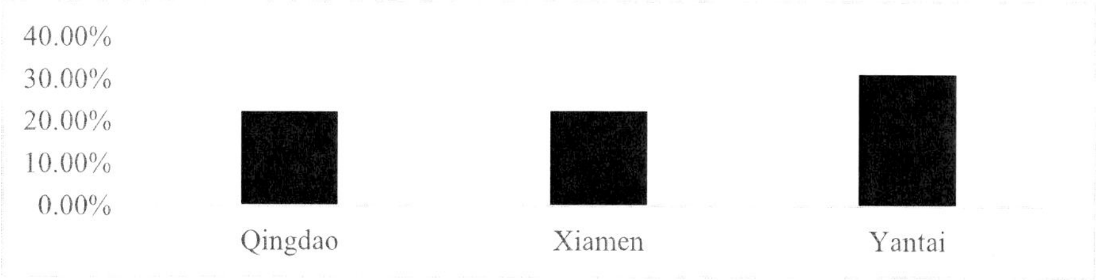

Fig. 4. Comparison of non-passenger working hours in different cities

6.2 Setting of Regulation Parameters

6.2.1 Annual Average Passenger Carrying Speed

In order to give consideration to the efficiency of enterprise transportation organization and the safety of driving, consider the construction objectives of transit metropolis and transit network optimization, and avoid reducing the average speed to increase the passenger carrying time and affecting the service quality, combined with the historical speed data of Yantai urban transit lines, this regulation takes 23 km/h as the regulation.

6.2.2 Ratio of Non-passenger Working Hours to Total Working Hours

Due to the shortage of manpower, the current scheduling efficiency is too low, resulting in long non passenger time. Therefore, while the driver's salary level is close to other cities, the work efficiency should also be gradually close to other cities. With reference to the proportion of other cities, the regulation value is 22.5%.

6.2.3 Reference Wage Benchmark

The average wage of non-private enterprises in cities and towns is chosen as the reference wage.

6.2.4 Department of Wage Growth Control

Design the salary growth control method which is linked with the operating cost control and service quality assessment results of public transport enterprises. When the performance appraisal results of public transport enterprise service quality are excellent, the adjustment coefficient of wage growth is 1.2; when the performance appraisal results of public transport enterprise service quality are qualified, the adjustment coefficient of wage growth is 1; when the performance appraisal results of public transport enterprise service quality are unqualified, the adjustment coefficient of wage growth is 0.

6.2.5 Parameter Setting Considering the Standard Working Hours and Salary Level of Drivers

The following factors should be considered when determining the annual standard working hours per capita and the driver's salary level:

①From the perspective of sustainable and stable development of the industry, the average wage of drivers should not be lower than the average local social wage;

②According to the industry standard, the annual standard working hours are recommended to be 1880–2432 h, and the value should be 1920 h; the wage coefficient should be 115–130% of the average social wage; according to Yantai's 2018 standard, that is, under the 1920 h working intensity, the driver's wage level should be 88%–105% of the average non private urban wage;

③According to the recommended value of work intensity of industry standard, the wage level of drivers in the same type of cities is compared horizontally. The average wage of bus drivers in Yantai is 1920 h, which only accounts for about 65% of the average wage of Urban Non private employees. There is still a big gap between the average wage of Yantai bus drivers and that of other cities (90%).

Based on the above factors, several sets of wage levels are designed as follows:

Among them, under the intensity of 1–3:1920 working hours (3844 approved), the wage level reaches the medium high level in other cities, that is, 96% of the average wage of Urban Non private employees on duty; 2432 h, 122% of the wage level, 2300 h and 115% of the wage level, which is actually different forms of the same total wage; the ratio of the total driver's salary to the average level of the previous three years It rose more than 129 per cent.

Scheme 4–6: under the intensity of 1920 working hours (authorized number of 3844 people), the wage level reaches the medium low level of other cities, that is, 88% of the average wage of Urban Non private employees; 2432 h, 111.5% of the wage level, 2300 h, 105% of the wage level are actually different forms of the same total wage; the total wage of drivers is higher than the average level of the previous three years More than 118% (Table 1).

Table 1. Different schemes of driver wage level and labor intensity parameter setting

		Annual standard working hours	**Approval of number of pilots**	**Ratio to the urban non-private sector in that year**	**Ratio to actual total wage**	
	The actual situation	3268	2506	112.00%		
Close to the average level of drivers in other cities	Option 1	1920	3844	96.00%	129%	
	Option 2	2432	3035	122.00%		
	Option 3	2300	3209	115.00%		
Close to the lower limit of industry standards	Option 4	1920	3844	88.00%	118%	

(continued)

Table 1. (*continued*)

		Annual standard working hours	**Approval of number of pilots**	**Ratio to the urban non-private sector in that year**	**Ratio to actual total wage**	
	The actual situation	3268	2506	112.00%		
	Option 5	2432	3035	111.50%		
	Option 6	2300	3209	105.00%		Recommended scheme√

After considering the financial affordability, the relevant provisions of the labor law, the need for sustainable development of the industry and the relatively acceptable number, the regulation recommended scheme 6: 2300 h of standard working hours in the whole year. When the performance appraisal is qualified, the ratio of the driver's wage level to the social non private wage of that year is 105%.

6.3 Regulation Method of Bus Driver's Total Wage

In summary, the formula for regulating the total wages of drivers in Yantai is as follows:

$$JZ = JR \times JG = \left[\frac{\frac{ZL}{S(23\text{Km/h})}}{\frac{1-f(22.5\%)}{Rt(2300\text{ hour})}} \right] \times [W \times (1 + Rg \times b) \times (1 + a)] \tag{5}$$

where, the same letters and values are given above.

The regulation method of bus driver's total salary has been applied to the regulation method of bus cost in the implementation plan of Yantai municipal government's purchase of public transport services, which is used for the budget and approval of Yantai's bus operation cost and service purchase cost from 2021 to 2023.

7 Concluding Remarks

At present, this regulation method has been applied in Yantai, and the application effect is good, which ensures the labor intensity and wage level of front-line drivers. In fact, for the regulation of bus operation cost, the total salary of drivers is only one of the labor costs. In addition, there are other labor costs such as management auxiliary personnel and wage expenses. In addition to labor costs, major cost items such as fuel and power consumption, warranty fees, tire fees, depreciation of fixed assets, leasing fees, insurance premiums and accident losses also need to be regulated by comprehensive design methods and parameters on the premise of considering labor intensity, work efficiency, financial affordability and other factors.

For the purchase of public transport services, cost regulation is only a basis to determine the rationality of the purchase of services. In addition, service standards, service quality assessment, ticket system and fare dynamic adjustment are also needed to maximize the efficiency of the use of funds for the purchase of services.

For a city's public transport, the government's purchase of public transport services is only an incentive way for self-improvement within the industry. In addition, no matter the priority development of public transport, the promotion of attraction, or the construction of transit city, it is neither the responsibility of the industry management department nor the task of one stage in the implementation period of the system, but a comprehensive, long-term and systematic project. It is a long-term task for all cities.

References

1. State Council, opinions of the State Council on the reform of wage determination mechanism in state owned enterprises (SFF [2018] No. 16) [EB / OL] (May 25, 2018)
2. Ministry of transport, specification for operation cost calculation of urban bus and tram enterprises (JTT1184–2018) [EB / OL] (May 1, 2018)
3. Wang, J., Zhong, Z., Yu, K.: Research on the design of urban public transport government purchase service system based on the linkage of cost, income and subsidy. Transp. Manager World **3**, 76–79 (2017)
4. Zhong, Z.: Theoretical exploration and promotion path of government purchasing public transport services. Transp. Eng. **6**, 51–57 (2018)
5. Wang, J., Xu, C.: Public transport operation service quality assessment mode design from extensive to refined. Transp. Manager World **4**, 86–91 (2017)

A Car-Following Model Considering the Next-Nearest-Neighbor Interaction of Electronic Throttle Information

Yirong Kang[1](✉) and Shuhong Yang[2]

[1] School of Transportation Engineering, Guizhou Institute of Technology, Guiyang 550003, China
kyr6700256@126.com
[2] School of Computer Science and Communication, Guangxi University of Science and Technology, Liuzhou 545006, China

Abstract. Aiming at the problem that the nearest preceding car of the considered vehicle may leave the current lane and cause the latter to lose track reference, based on the classical FVD model, this paper design a more effective feedback control signal by considering the ET opening angle difference information from the next-nearest-neighbor interaction at the previous moment, and propose a new CAV cooperative driving following model. The effect of this new consideration upon the stability of CAV traffic flow is examined through linear stability analysis. A modified Korteweg-de Vries (mKdV) equation was derived via nonlinear analysis to describe the propagating behavior of traffic density wave near the critical point. Good agreement between the simulation and the analytical results shows that taking the next-nearest-neighbor interaction of electronic throttle historical information into account leads to the stabilization of traffic systems, and thus can efficiently suppress the emergence of traffic jamming. The results can be expected to provide a theoretical reference for designing more effective traffic congestion control strategies.

Keywords: Car-following Model · Electronic Throttle · Next-nearest-neighbor Interaction · Connected and Autonomous Vehicle

1 Introduction

With the fast growth of the economy and the pace of social development, the problem of traffic congestion is becoming increasingly prominent, which seriously hinders people's daily work and life. Studying on alleviating traffic congestion has attracted extensive attention, and many traffic flow models [1–5] have been proposed to provide theoretical support for solving traffic problems. At present, the existing traffic flow models can be roughly divided into three categories: macroscopic models, microscopic models, and mesoscopic models. The car-following model is a favorable type of microscopic traffic models, which describe the interaction between each pair of leading and following vehicles in the same lane based on the formula of stimulus response framework.

X. Zeng et al. (Eds.): ITASC 2022, LNEE 1042, pp. 157–168, 2023.
https://doi.org/10.1007/978-981-99-2252-9_15

Among them, the most well-known one is the optimal velocity (OV) model proposed by Bando et al. [4], which has successfully revealed the dynamical evolvement process of traffic congestion in a simple way. Hereafter, various OV-based car-following models have been proposed one after another by introducing actual traffic factors, such as multiple information of headway [5] or relative velocity [6], backward-looking effect [7], driver's anticipation [8], curved road condition [9] and so on. These car-following models mentioned above can reproduce complex traffic phenomena and reveal mechanism of traffic jams at a higher level.

Recently, along with the Vehicle-Vehicle communication technology, modeling the car following process of Connected and Autonomous Vehicle (CAV) has attracted increasing attention in the field of traffic flow. In this direction, some scholars successively developed some extended car following model by factoring the effect of electronic throttle (ET) opening angle. In 2016, based on the full velocity difference (FVD) model, Li et al. [10] proposed a throttle-based car following model (recorded as T-FVD) to describe the effect of ET opening angle on CAV traffic flow. After that, considering the factors of lateral gap and electronic throttle opening angle in the same time, they presented an extended car-following model under connected environment in 2017 [11]. Their results show that the stability of traffic flow with CAV is improved by introducing the new consideration. By making full use of ET angle control signal of multiple preceding vehicles, a novel car-following model was established with consideration of optimal velocity difference and ET angle information, in which the time-dependent Ginzburg-Landau (TDGL) equation and the modified Korteweg-de Vries (mKdV) equation are inferred to describe the evolutionary process of density wave [12]. In 2019, Sun et al. proposed an extended car-following model by taking the effect of ET signal into account on the curved road [13], in which the ET opening angle difference from multiple preceding vehicles at the previous moment is considered as a delay-feedback control signal.

A series of research results of CAV car-following model mentioned above demonstrate that with the help of vehicle-vehicle communication technology, the traffic jams can be effectively suppressed by exchanging the ET opening angle data among continuous preceding vehicles. However, the existing CAV car-following models are unsuited to study the ET opening angle influence of the next-nearest-neighbor interaction (non-adjacent the current vehicle), since they do not consider this factor at all. In fact, classic traffic flow research results [14, 15] have proved that the next-nearest-neighbor interaction factors have an important influence on car following scene. Naturally, there is an open question regarding to whether or not the next-nearest-neighbor interaction in electronic throttle dynamics affects the CAV car-following. However, to our knowledge, this open question has not been explored in the CAV car-following model up to now. In view of the above reason, in this paper, a new car-following model is constructed by considering the effect of ET dynamics for CAV traffic flow, in which the ET opening angle difference from the next-nearest-neighbor interaction at the previous moment is considered in the vehicle networking environment. The analytical and numerical simulation are conducted to show that the new consideration has significant influence on the stability and dynamic characteristics of CAV traffic flow.

2 Models

In 1995, Bando et al. proposed the OV model [4] to describe car-following behavior on a single-lane highway. The motion equation is as follows:

$$\frac{dv_j(t)}{dt} = a[V(\Delta x_j(t)) - v_j(t)] \tag{1}$$

where $x_j(t)$ and $v_j(t)$ are the position and velocity of the *jth* car respectively, and t represents time. a denotes the sensitivity of the driver and is given by the inverse of the delay time τ, namely $a = 1/\tau$. $\Delta x_j(t) = x_{j+1}(t) - x_j(t)$, represents the headway of two successive vehicles. $V(\cdot)$ is the optimal velocity function. Comparison with empirical data shows that too high acceleration and unrealistic deceleration occurs in OV model.

To overcome the deficiency of OV model, Helbing and Tilch [16] proposed a generalized force (GF) model via introducing a negative velocity difference into the OV model. Later, considering the positive relative velocity into the GF model, Jiang et al. [17] developed the full velocity difference (FVD) model as follows:

$$\frac{dv_j(t)}{dt} = a[V(\Delta x_j(t)) - v_j(t)] + \lambda \Delta v_j(t) \tag{2}$$

where $\Delta v_j(t) = v_{j+1}(t) - v_j(t)$ is the velocity difference between the leading car $j + 1$ and the following car j at time t. λ is the responding factor of the velocity difference. The results illustrate that the FVD model has better agreement with field data, compared with the OV and GF models. However, unrealistically high deceleration also occurs in the FVD model [17].

To capture the nature of CAV traffic flow, Li et al. [10] proposed the throttle-based FVD (T-FVD) model by introducing the effect of ET opening angle into the FVD model. The dynamic equation is described as follows:

$$\frac{dv_j(t)}{dt} = a[V(\Delta x_j(t)) - v_j(t)] + \lambda \Delta v_j(t) + \gamma \Delta \theta_j(t) \tag{3}$$

where $\theta_j(t)$ is the electronic throttle opening angle of *jth* vehicle at time t. $\Delta\theta_j(t) = \theta_{j+1}(t) - \theta_j(t)$ represents the ET opening angle difference between the vehicle $j + 1$ and the vehicle j at time t. $\gamma > 0$ is the control coefficient of the angle difference.

According to the existing literature [10, 12], the mathematical relationship between the electronic throttle angle and the velocity difference and acceleration can be expressed as follows.

$$\frac{dv_j(t)}{dt} = -p(v_j(t) - v_e) + q(\theta_j(t) - \theta_e) \tag{4}$$

where p and q are constant greater than zero, v_e is the current equilibrium speed, θ_e is the electronic throttle angle, it corresponds to the current equilibrium speed.

According to the idea mentioned in the introduction, this paper proposes an extended model by considering the information of the ET opening angle difference from the next-nearest-neighbor interaction at the previous moment. The proposed model is represented as follows:

$$\frac{dv_j(t)}{dt} = a[V(\Delta x_j(t)) - v_j(t)] + \lambda \Delta v_j(t) + k[\theta_{j+2}(t-1) - \theta_{j+1}(t-1)] \tag{5}$$

where the next-nearest-neighbor interaction term $\theta_{j+2}(t-1)-\theta_{j+1}(t-1)$ is the historical information of ET opening angle difference between the $(j+2)th$ vehicle and its following vehicle $(j+1)th$ at time $t-1$. k is the sensitivity coefficient. The control strategy of this extended CAV car following model is that the host car adjusts his acceleration not only by the received information of velocity $v_j(t)$ and velocity difference $\Delta v_j(t)$ of the jth car at time t, but also by the collected signal of ET opening angle difference $\theta_{j+2}(t-1)-\theta_{j+1}(t-1)$ from the next-nearest-neighbor interaction at the previous moment $t-1$. When $k=0$, the new model degenerates into FVD model [17].

According to Eq. (4), we can get

$$\theta_j(t) = \frac{1}{q}\left[\frac{dv_j(t)}{dt} + p(v_j(t) - v_e)\right] + \theta_e \tag{6}$$

Thus, we can obtain the opening angle difference of electronic throttle between the $(n+2)$ th and the $(n+1)$ th vehicles at time t-1 as

$$\theta_{j+2}(t-1) - \theta_{j+1}(t-1) = \frac{1}{q}\Big[\frac{dv_{j+2}(t-1)}{dt} - \frac{dv_{j+1}(t-1)}{dt} + p(v_{j+2}(t-1) - v_{j+1}(t-1))\Big] \tag{7}$$

Substituting Eq. (7) into Eq. (5), we can rewrite Eq. (5) as:

$$\frac{dv_j(t)}{dt} = a[V(\Delta x_j(t)) - v_j(t)] + \lambda \Delta v_j(t) + k\cdot\frac{1}{q}\left[\frac{dv_{j+2}(t-1)}{dt} - \frac{dv_{j+1}(t-1)}{dt} + p(v_{j+2}(t-1) - v_{j+1}(t-1))\right] \tag{8}$$

For the convenience of nonlinear analysis, Eq. (8) can be further rewritten as:

$$\frac{d^2\Delta x_j(t)}{dt^2} = a[V(\Delta x_{j+1}(t)) - V(\Delta x_j(t)) - \frac{d\Delta x_j(t)}{dt}] + \lambda[\frac{d\Delta x_{j+1}(t)}{dt} - \frac{d\Delta x_j(t)}{dt}] + k\cdot\frac{1}{q}\{\frac{d^2\Delta x_{j+2}(t-1)}{dt^2} - \frac{d^2\Delta x_{j+1}(t-1)}{dt^2} + p[\frac{d\Delta x_{j+2}(t-1)}{dt} - \frac{d\Delta x_{j+1}(t-1)}{dt}]\} \tag{9}$$

In this paper, we take the following optimal velocity function(for short,OVF) calibrated with the empirical data by Helbing [16]:

$$V(\Delta x_j) = V_1 + V_2 \tanh[C_1(\Delta x - l_c) - C_2] \tag{10}$$

Here, parameters in the OVF are set as $a = 0.85\ \text{s}^{-1}$, $V_1 = 6.75\ \text{m/s}$, $C_2 = 1.57$, $C_1 = 0.13\ \text{m}^{-1}$, $V_2 = 7.91\ \text{m/s}$ and the length of the vehicle is $l_c = 5$ m.

3 Linear Stability Analysis

There is no doubt that the performance of stability is the most important characteristic of CAV traffic flow control system. In this section, the linear stability analysis is conducted to investigate the stabilizing ability of ET in CAV scene. It is obvious that the CAV traffic flow reach steady state when the cars run with the uniform headway b and optimal velocity $V(b)$. Therefore, the steady-state solution is given as,

$$x_j^0(t) = bj + V(b)t,\ b = L/N \tag{11}$$

where L is the road length and N is the total car number. Suppose $y_j(t)$ is a small deviation from the steady state $x_j^0(t)$: $x_j(t) = x_j^0(t) + y_j(t)$. Substituting it into Eq. (8) and linearizing them yields

$$\begin{aligned} y_j''(t) &= a[V'(b)\Delta y_j(t) - y_j'(t)] + \lambda[y_{j+1}'(t) - y_j'(t)] \\ &\quad + k \cdot \frac{1}{q}\{y_{j+2}''(t-1) - y_{j+1}''(t-1) + p[y_{j+2}'(t-1) - y_{j+1}'(t-1)]\} \end{aligned} \tag{12}$$

where $V' = dV(\Delta x_j)/d\Delta x_j\big|_{\Delta x_j = b}$ And $\Delta y_j(t) \equiv y_{j+1} - y_j$.

Via expanding $y_j(t) = Ae^{ikj+zt}$,where $z = z_1(ik) + z_2(ik)^2 + \cdots$, and inserting them into Eq. (12). One has the first- and second-order terms of ik respectively,

$$z_1 = V'(b),\ z_2 = \frac{1}{2}V'(b) + (\frac{\lambda}{a} + \frac{1}{aq}kp)V'(b) - \frac{1}{a}[V'(b)]^2 \tag{13}$$

If $z_2 < 0$, the uniformly steady-state flow becomes unstable, while the uniform flow is stable when $z_2 > 0$. Thus the neutral stable criteria for this steady state is given by

$$a = 2[V'(b) - \lambda - \frac{1}{q}kp] \tag{14}$$

For small disturbance with long wavelengths, the homogeneous traffic flow is stable in the condition that

$$a > 2[V'(b) - \lambda - \frac{1}{q}kp] \tag{15}$$

As $k = 0$, the result of stable condition is the same as that of the FVD model [17].

$$a > 2[V'(b) - \lambda] \tag{16}$$

The neutral stable condition Eq. (14) clearly show that the sensitivity coefficient k play an important role in stabilizing the CAV traffic flow. The neutral stable curves under different values of k are depicted in the headway-sensitivity space $(\Delta x, a)$ for the proposed model with $p = 0.8$, $q = 0.27$, $\lambda = 0.3$. In Fig. 1, the headway-sensitivity space is divided into two different parts by each curve. The upper part of neutral stability line denotes the stable area, and the lower part of the curve represents unstable area. Different regions mean that traffic flow of CAV will show different evolution states when small disturbances are added. Generally, when the traffic flow is in a stable area, the small disturbances will eventually evolve into a uniform state of motion. However, when it is in an unstable state, the unstable signal will be gradually enlarged, and the corresponding traffic flow will eventually evolve into a congestion state with time. From Fig. 1, one can see that every curve exists apex (h_c, a_c) named critical point. As k increases, the unstable region is compressed gradually, and the corresponding critical point of each curve decreases gradually, which means that the new consideration leads to the stabilization of CAV traffic systems. Particularly, as $k = 0$, the neutral stability line is the same as that of FVD model [17].

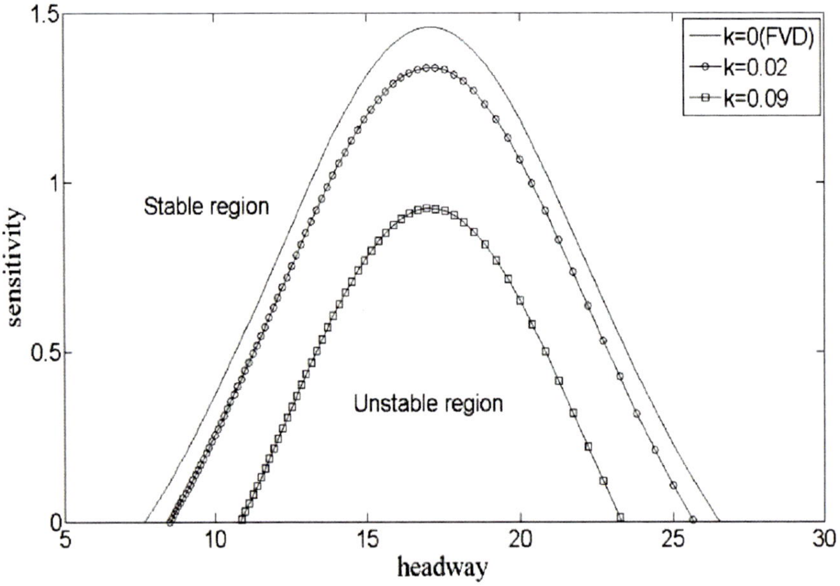

Fig. 1. The neutral stability curves in headway-sensitivity space $(\Delta x, a)$

4 Nonlinear Analysis and mKdV Equation

To investigate the slowly varying behavior of CAV traffic flow near the critical point (h_c, a_c), nonlinear analysis and derivation of the mKdV density wave equation are conducted to describe traffic jam propagation near critical point. For extracting slow scales with the space variable j and the time variable t, the slow variable X and T are defined as follows:

$$X = \varepsilon(j + bt) \text{ and}$$
$$T = \varepsilon^3 t, 0 < \varepsilon \leq 1 \tag{17}$$

where b is a constant to be determined. Given

$$\Delta x_j(t) = h_c + \varepsilon R(X, T) \tag{18}$$

Substituting Eq. (17) and Eq. (18) into Eq. (9) and making the Taylor expansion to the fifth order of ε lead to the following expression:

$$\begin{aligned} &\varepsilon^2(b - V')\partial_X R + \varepsilon^3\left(\frac{b^2}{a} - \frac{V'}{2} - \frac{\lambda b}{a} - \frac{1}{aq}kpb\right)\partial_X^2 R + \varepsilon^4\{\partial_T R - [\frac{V'}{6} + \frac{b\lambda}{2a} \\ &+ \frac{2kb^2 + kpb(3-2b)}{2aq}]\partial_X^3 R - \frac{1}{6}V'''\partial_X R^3\} + \varepsilon^5\{(\frac{2b}{a} - \frac{\lambda}{a} - \frac{1}{aq}kp)\partial_T\partial_x R \\ &-[\frac{V'}{24} + \frac{\lambda b}{6a} + \frac{3kb^2(3-2b) + kpb(3b^2 - 9b + 7)}{6aq}]\partial_X^4 R - \frac{1}{12}V'''\partial_X^2 R^3\} = 0 \end{aligned} \tag{19}$$

where $V' = [dV(\Delta x_j) \big/ d\Delta x_j]|_{\Delta x_j = h_c}$ and $V''' = [d^3V(\Delta x_j) \Big/ d\Delta x_j^3]|_{\Delta x_j = h_c}$.

Near the critical point (h_c, a_c), $\tau = (1+\varepsilon^2)\tau_c$, taking $b = V'$ and eliminating the second order and third order terms of ε from Eq. (19) result in the simplified equation:

$$\varepsilon^4[\partial_T R - g_1\partial_X^3 R + g_2\partial_X R^3] + \varepsilon^5[g_3\partial_X^2 R + g_4\partial_X^4 R + g_5\partial_X^2 R^3] = 0 \tag{20}$$

where

$$g_1 = \frac{V'}{6} + \frac{b\lambda\tau_c}{2} + \frac{2kb^2 + kpb(3-2b)}{2q}\tau_c \tag{21}$$

$$g_2 = -\frac{V'''}{6} g_2 = -\frac{V'''}{6} \tag{22}$$

$$g_3 = (b - \lambda - \frac{1}{q}kp)b\tau_c \tag{23}$$

$$\begin{aligned} g_4 = (2b - \lambda - \frac{1}{q}kp)\tau_c[\frac{V'}{6} + \frac{b\lambda\tau_c}{2} + \frac{2kb^2 + kpb(3-2b)}{2q}\tau_c] \\ -[\frac{V'}{24} + \frac{\lambda b\tau_c}{6} + \frac{3kb^2(3-2b) + kpb(3b^2 - 9b + 7)}{6q}\tau_c] \end{aligned} \tag{24}$$

$$g_5 = \frac{1}{12}[(4b - 2\lambda - \frac{2}{q}kp)\tau_c - 1]V''' \tag{25}$$

To derive the regularized equation, the following transformations are performed on Eq. (20):

$$T' = g_1 T,\ R = \sqrt{\frac{g_1}{g_2}}R' \tag{26}$$

The standard mKdV equation with a $O(\varepsilon)$ correction term is given as follows:

$$\partial_{T'}R' - \partial_X^3 R' + \partial_X R'^3 + \varepsilon M[R'] = 0 \tag{27}$$

where

$$M[R'] = \sqrt{\frac{1}{g_1}[g_3\partial_X^2 R' + g_4\partial_X^4 R' + \frac{g_1 g_5}{g_2}\partial_X^2 R'^3]} \tag{28}$$

Ignoring the $O(\varepsilon)$ term, we can obtain the mKdV equation with the kink-antikink soliton solution:

$$R_0'(X, T') = \sqrt{c}\tanh\sqrt{\frac{c}{2}}(X - cT') \tag{29}$$

With the method described in Ref. [18], we obtain the selected velocity c.

$$C = \frac{5g_2 g_3}{2g_2 g_4 - 3g_1 g_5} \tag{30}$$

Hence, we obtain the kink-antikink soliton solution as follows:

$$\Delta x_j(t) = h_c + \sqrt{\frac{g_1 C}{g_2}(\frac{\tau}{\tau_c} - 1)} \tanh \sqrt{\frac{C}{2}(\frac{\tau}{\tau_c} - 1)}[j + (1 - Cg_1(\frac{\tau}{\tau_c} - 1))t] \quad (31)$$

Then, amplitude A of the kink-antikink soliton is given by

$$A = \sqrt{\frac{g_1 c}{g_2}(\frac{\tau}{\tau_c} - 1)} \quad (32)$$

The kink-antikink solution represents the coexisting phases, which consists of the freely moving phase at low density and the congestion jam at high density. Through nonlinear analysis, the propagating property of traffic jams near the critical point can be described by the solution of the mKdV density wave equation. Based on Eq. (30), the values of propagation velocity C under different values of k are shown in Table 1 for the proposed model. From Table 1, it can be found when k increases, the critical value a_c and the propagation speed C of traffic waves (jams) decrease gradually, which means that the stability of traffic flow increases gradually and the traffic jams has been reduced significantly by considering the effect of ET opening angle difference from the next-nearest-neighbor interaction at the previous moment.When $k = 0$, the result is the same as that in the FVD model [17].

Table 1. The critical sensitivity a_c and the propagation speed C for the proposed model when $p = 0.8, q = 0.27, \lambda = 0.3$.

K	0	0.02	0.04	0.06	0.08
a_c	1.4566	1.3973	1.3381	1.2788	1.2196
C	2.3376	1.7415	1.3288	1.0271	0.7977

5 Numerical Simulation and Validation

The numerical simulation is carried out under a periodic boundary condition to verify the effectiveness of theoretical analysis for the new model. The platform for all numerical simulation and validation is a computer with Intel Core(TM) i7 8700 CPU @ 3.2 GHz and 16G RAM. Matlab is employed as the programming language. The total car number $N = 100$ and circuit length $L = 1500$ m. The related parameters are taken as $a = 1$, $p = 0.8$, $q = 0.27$, $\lambda = 0.3$. The initial disturbance is same as that in Ref. [18]:

$$x_1(0) = 1\,m, x_1(0) = (n-1)L/N\,m \text{ for, n} \neq 1 \quad (33)$$

$$\dot{x}_n(0) = V(L/N)\,(n = 1, 2, \cdots, N) \quad (34)$$

Figures 2 and 3 demonstrate snapshots of all vehicles' velocity and headway at $t = 1000s$ for different values of k respectively. When $k = 0$, the results are consistent with FVD model [17]. In Fig. 2 and 3, when $k = 0$ and $k = 0.02$, the stability conditions are not satisfied, the initial disturbance propagates along the upstream of the CAV traffic flow and enlarges gradually, and finally evolves into a stop-and-go waves (traffic jams).However, under the same condition, because of the new consideration is introduced into the proposed model, one can observe clearly in Fig. 2 and Fig. 3 that the fluctuation amplitude of speed and headway of the new model are smaller than that of FVD model [17], which means that the stability is improved by incorporating the historical information of next-nearest-neighbor interaction of electronic throttle. Furthermore, with the increasing value of k, the number of stop-and-go waves and the corresponding fluctuation amplitude reduced obviously in the snapshot of velocity and headway, hence the system stability of CAV traffic flow is enhanced effectively. Especially, when $k = 0.09$, the stability condition hold true and thus the initial small disturbance signal will gradually decay, and the CAV traffic flow finally become stable again after a sufficiently long time.

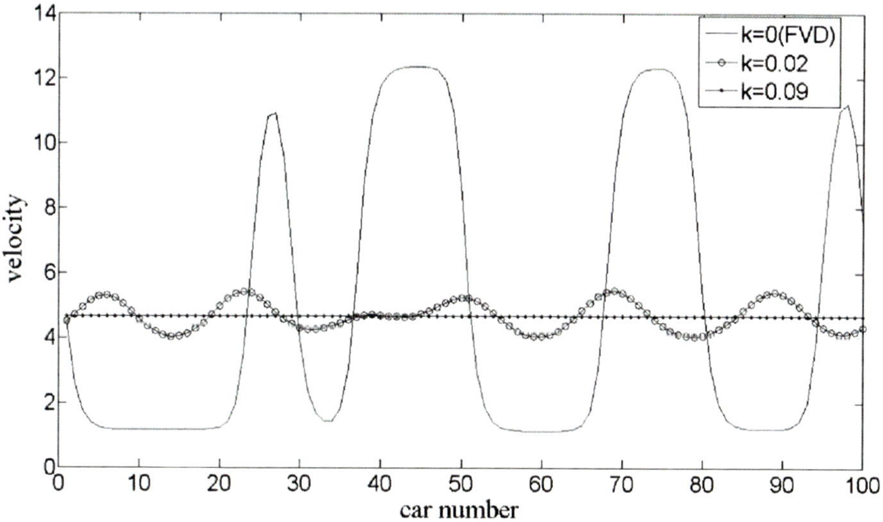

Fig. 2. Snapshot of the velocities of all vehicles at different values k when $t = 1000s$ in the new model ($a = 1, p = 0.8, q = 0.27, \lambda = 0.3$).

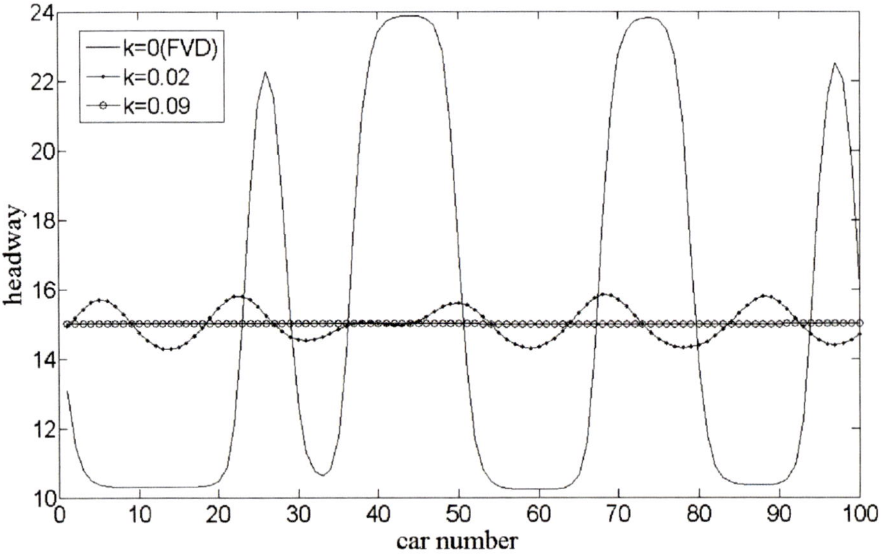

Fig. 3. Snapshot of the headway of all vehicles at different values k when $t = 1000s$ in the new model ($a = 1, p = 0.8, q = 0.27, \lambda = 0.3$).

Figure 4 shows the velocity-headway trajectory of the proposed model for different value of k. After sufficiently long time, the closed trajectories called hysteresis loops are clearly observed. In Fig. 4, as the value of k increases, the area of hysteresis loop becomes progressively smaller. This indicates that the new consideration plays a positive function on the stabilization of traffic flow. Particularly, when $k = 0.09$, the loop will shrink into a point on the optimal velocity curve, and the traffic flow get the stability state.

In order to investigate the performance of the proposed model further, we analyze the start-up velocity and delay time of vehicle queue compared with classical FVD model. The initial conditions and parameters are consistent with reference [17]: When $t < 0$, the signal light is red, the 11 cars are evenly lined up at intervals of 7.4 m, the corresponding speed is zero, and when t = 0, the signal light changes from red to green, all the cars began to speed up.

We define the delay time of car motion by δt as that in FVD model [17], Then the kinematic wave speed can be defined as $c_j = 7.4/\delta t$.The statistical results are shown Table 2 by taking the same parameters as those in FVD model [17].

As shown in Table 2, the velocity c_j of the proposed model are included in the expected range [17 km/h, 23 km/h] just as Bando et al. [4] pointed out, while the delay time δt is obviously shorter than FVD model. Which illustrates that by considering the ET opening angle difference information from the next-nearest-neighbor interaction at the previous moment, our model can respond to the change of preceding car more quickly, and thus is benefit to the smoothing of traffic.

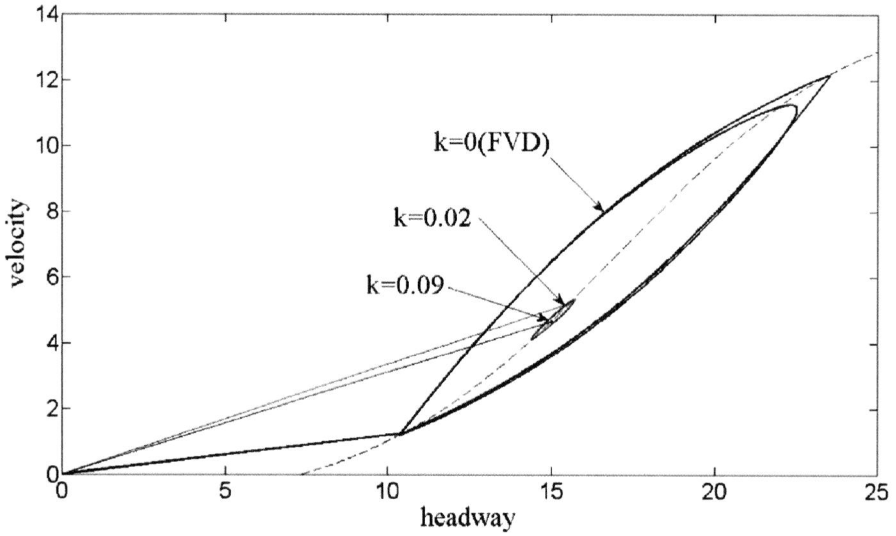

Fig. 4. The loops for new model under different values k. The dashed curve represents OVF.

Table 2. δt and c_j in FVD model and the proposed model.

Model	$a(s^{-1})$	$\lambda(s^{-1})$	$k(s)$	$\delta t(s)$	$c_j(km/h)$
FVD	0.41	0.5	0	1.4	19.03
Our model	0.41	0.5	0.01	1.2	22.11

6 Conclusions

In this paper, a new car-following model is proposed by considering the information of next-nearest-neighbor interaction of electronic throttle at the previous moment for the connected Autonomous Vehicles environment. The influence of the electronic throttle information on the stability of CAV traffic flow is investigated by using the linear stability theory. Furthermore, on the basis of previous linear analysis, the mKdV equation has been derived to describe the density wave of CAV traffic flow near the critical point through nonlinear analysis. The simulation results are in good agreement with theoretical analysis and further indicate that, taking the historical information of next-nearest-neighbor interaction of electronic throttle into account, the traffic congestion can be effectively alleviated and the stabilization of CAV car following process can be achieved. Therefore, it is reasonable to conclude that historical information of the next-nearest-neighbor interaction of electronic throttle plays an important role in stabilizing the traffic flow in car following process and thus should be considered in traffic flow modeling.

Acknowledgments. This work is supported by Science and Technology Plan Projects of Guizhou Province: (No. 20181059) and Natural Science Foundation of Guangxi (No. 2018GXNS-FAA050020).

References

1. Sun, B., Wu, N., Ge, Y.E., et al.: A new car following model considering acceleration of lead vehicle. Transport **31**(1), 1–10 (2016)
2. Zhang, Y., Ni, P., Li, M., et al.: A new car-following model considering driving characteristics and preceding vehicle's acceleration. J. Adv. Transp. 1–14 (2017)
3. Cao, B.G.: A car-following dynamic model with headway memory and evolution trend. Physica A **539**, 122903 (2020)
4. Bando, M., Hasebe, K., Nakayama, A., et al.: Dynamical model of traffic congestion and numerical simulation. Phys. Rev. E 51(2), 1035 (1995)
5. Redhu, P., Siwach, V.: An extended lattice model accounting for traffic jerk. Physica A **492**, 1473–1480 (2018)
6. Ge, H.X., Cheng, R.J., Li, Z.P.: Two velocity difference model for a car following theory. Physica A **387**, 5239–5245 (2008)
7. Ge, H.X., Zhu, H.B., Dai, S.Q.: Effect of looking backward on traffic flow in a cooperative driving car following model. Eur Phys J B **54**, 503–507 (2006)
8. Tang, T.Q., Li, C.Y., Huang, H.J.: A new car-following model with the consideration of the driver's forecast effect. Phys. Lett. A **374**(38), 3951–3956 (2010)
9. Zhang, L.D., Jia, L. Zhu, W.X.: Curved road traffic flow car-following model and stability analysis. Acta Phys. Sin. **61**(7), 074501 (2012)
10. Li, Y.F., Zhang, L., Peeta, S., et al.: A car-following model considering the effect of electronic throttle opening angle under connected environment. Nonlinear Dynam. **85**, 2115–2125 (2016)
11. Li, Y.F., Zhao, H., Zheng, T.X., et al.: Non-lane-discipline-based car-following model incorporating the electronic throttle dynamics under connected environment. Nonlinear Dynam. **90**, 2345–2358 (2017)
12. Yan, C.Y., Ge, H.X., Cheng, R.J.: An extended car-following model by considering the optimal velocity difference and electronic throttle angle. Physica A **535**, 122216 (2019)
13. Sun, Y.Q., Ge, H.X., Cheng, R.J.: A car-following model considering the effect of electronic throttle opening angle over the curved road. Physica A **534**, 122377 (2019)
14. Nagatani, T.: Stabilization and enhancement of traffic flow by the next-nearest-neighbor interaction. Phy. Rev. E **60**(6), 6395–6401 (1999)
15. Li, Z.P., Liu, Y.C., Liu, F.Q.: A dynamical model with next-nearest-neighbor interaction in relative velocity. Int. J. Mod. Phys. C **18**(05), 819–832 (2011)
16. Helbing, D., Tilch, B.: Generalized force model of tragic dynamics. Phys. Rev. E **58**, 133–138 (1998)
17. Jiang, R., Wu, Q.S., Zhu, Z.J.: Full velocity difference model for a car-following theory. Phys. Rev. E **64**, 017101 (2001)
18. Ge, H.X., Cheng, R.J., Dai, S.Q.: KdV and Kink-Antikink solitons in car-following models. Physica A **357**(3–4), 466–476 (2005)

A Speed Measurement Method of Rail Transit Based on Millimeter-Wave Radar

Tuo Shen[1,2], Lanxin Xie[1(✉)], Jinnan Luo[3], and Xiaoqing Zeng[1]

[1] Shanghai Key Laboratory of Rail Infrastructure Durability and System Safety, Shanghai 201804, China
918003582@qq.com

[2] School of Optical-Electrical and Computer Engineering, University of Shanghai for Science and Technology, Shanghai 200093, China

[3] Institute of Rail Transit, Tongji University, Shanghai, China

Abstract. The growing use of millimeter-wave radars in the intelligent transportation field and increasing accuracy provide new possibilities for train speed measurement. This paper proposes a measuring train speed method by using millimeter-wave radar to detect static objects. The millimeter-wave radar was used to obtain and process objects data. Then, using the RANSAC algorithm to extract the static objects group, and calculated the relative speed of the millimeter-wave radar and the static objects group. Finally, we completed the comparative experiment of using millimeter-wave radar and odometer to measure train speed in a real track environment. The results show that the measurement error of this method is less than 1.5 m/s, and it has the characteristics of low cost, fast speed measurement, and high precision.

Keywords: Rail transit · Train operation control · Speed measurement · Radar · Object detection

1 Introduction

With the development and application of new technologies such as cloud computing and artificial intelligence, rail transit has begun to move towards an intelligent development stage, which has higher requirements for the real-time and accuracy of train speed and position information provided by train speed and distance measurement equipment. Now, the more mature methods used in train speed measurement include wheel and axle speed sensor speed measurement, visual sensor-based speed measurement, satellite navigation-based speed measurement and positioning, and Doppler radar speed measurement.

(1) Odometer sensor

Odometer sensors are widely used in urban rail transit because of their economy, practicality, and mature technology, which can provide high-precision, digital speed, and distance information. However, this method uses wheel rotation as the collection

X. Zeng et al. (Eds.): ITASC 2022, LNEE 1042, pp. 169–181, 2023.
https://doi.org/10.1007/978-981-99-2252-9_16

object to indirectly obtain the train speed. Wheel diameter changes caused by wheel wear, idling,and sliding during operation will produce great measurement errors [1]. Liu Jiang et al. [2] analyzed the error characteristics of the wheel axis speed sensor in the train combined positioning and proposed a wheel diameter calibration method based on gray theory based on the GPS-assisted wheel diameter calibration. But the shortcomings of this method are also obvious, it cannot provide speed correction, and the construction and maintenance costs are high.

(2) **Video sensor**

With the rapid development of digital image processing technology, video processing technology is gradually being applied to traffic detection. Vision sensors use one or more camera image sequences to estimate vehicle speed. However, the monitoring range of this speed measurement method is limited and it cannot adapt to the changing diversity of the environment. Badino H [3] proposes to use the features detected by the two cameras to match between the two images to obtain three-dimensional points, and then use the optical flow method to calculate the motion. Dahl Mattias et al. [3] in 2016 proposed a method of further calculating the vehicle speed by using the frame rate of the video collected by the road surveillance camera and the vehicle detection line to estimate the probability density function of the vehicle speed. However, in rail transit, due to the characteristics of high-speed trains, the effects of video images collected by video capture equipment are poor, and the accuracy of speed measurement is insufficient, which cannot meet the control requirements of the autonomous train operation control system. Therefore, video-based train speed measurement has almost no application.

(3) **Satellite navigation system**

The GNSS satellite navigation and positioning system represented by BDS can understand the train's location and speed in real-time. The principle is that the satellite continuously sends a unique code sequence that can be tracked. After the user receives this information, the user can calculate the three-dimensional position, three-dimensional motion speed and direction, and time information of the receiver [5]. However, the satellite signal is blocked or blocked in the space propagation process, which leads to limited availability of positioning functions. In scenarios such as tunnels and mountainous terrain, the availability of satellite positioning is very low. Ales [6] also demonstrated that a single GPS receiver cannot meet the safety requirements of rail transit.

(4) **Radar sensor**

In recent years, radar technology has gradually been applied to daily life in the military field. The principle of speed measurement is: First, the Doppler effect is used as the basic theory, and then the speed is calculated by the frequency signal between the echo and the transmitted wave received by the radio frequency module during the operation of the train [7]. Radar has the advantages of low environmental impact and high accuracy. Its value is similar to the actual speed of the train [8], and it is currently the mainstream speed measurement method. Recent studies based on radar speed measurement are all based on the Doppler effect to calculate train speed [9, 10]. Zhou Y [11] proposed a double-radar speed measurement calculation method in rail transit, which optimizes the measurement accuracy and improves the system reliability, but the cost is higher. Speed measurement methods using radar are becoming more and more feasible now.

The following Table 1 analyzes the characteristics of several mature speed sensors.

Table 1. Comparison of advantages and disadvantages of speed measurement methods

Speed Measurement Method	Advantages	Disadvantages
Odometer Sensor	Simple structure, mature technology, wide application	Error caused by idling and train sliding, low reliability
Vision sensor	Strong stability and low cost	Poor performance in a high-speed operation environment
Satellite navigation system	No train idling and sliding errors	Great environmental impact and blind spots
Radar sensor	Lightweight equipment, high environmental adaptability	Restricted by algorithm and device

The frequency band of the millimeter-wave is higher than that of radio. It has the characteristics of high accuracy, strong anti-interference ability, long detection range, wide detection range, etc. It can detect the relative distance, azimuth, and relative speed of the vehicle and the object in front of [12]. It has good application value in speed measurement and distance measurement. Therefore, this research proposes a method for measuring train speed using static objects detected by millimeter-wave radar, which is suitable for any type of radar that can classify objects as static and has universal applicability.

2 Design of Train Speed Measurement Algorithm

For each measurement cycle, perform the following steps. First, perform data cleaning on the measurement data of the millimeter-wave radar to filter out invalid object points. The premise of the experiment is that millimeter-wave radar data is measured in a non-moving environment, so the point cloud data is mainly static objects. Then extract the largest object group with the same speed value from it. If there is no object group with more objects. It is allowed to classify all the objects of the largest group found as static objects. The speed of the radar sensor can be reconstructed by the return speed of the static object. Finally, the train speed can be calculated by the conversion of the radar and the train motion coordinate system (see Fig. 1).

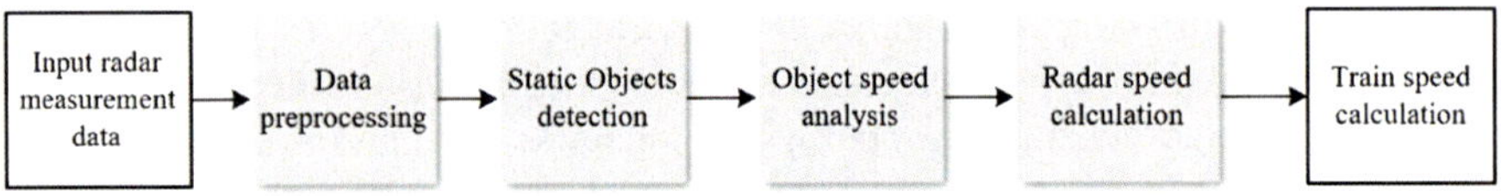

Fig. 1. The flow chart of the train speed measurement algorithm

Based on the analysis of the movement characteristics of the objects in front of the train and the detection mechanism of the on-board radar information, and considering the characteristics of the ground train movement is mainly two-dimensional movement and low mobility. We establish a train movement model based on the two-dimensional coordinate system of the train movement.

2.1 Coordinate Transformation

A. Establish radar coordinate system

The scanning plane of millimeter-wave radar is two-dimensional, and the original data obtained by scanning is a polar coordinate system based on distance and angle (see Fig. 2). To facilitate subsequent calculations, it needs to be converted into a rectangular coordinate system. The center position of the radar coordinate system is the installation position of the radar. The radar sensor processes the reflected signal and displays it in the form of "objects"after multiple steps, indicating the historical trajectory of the objects.

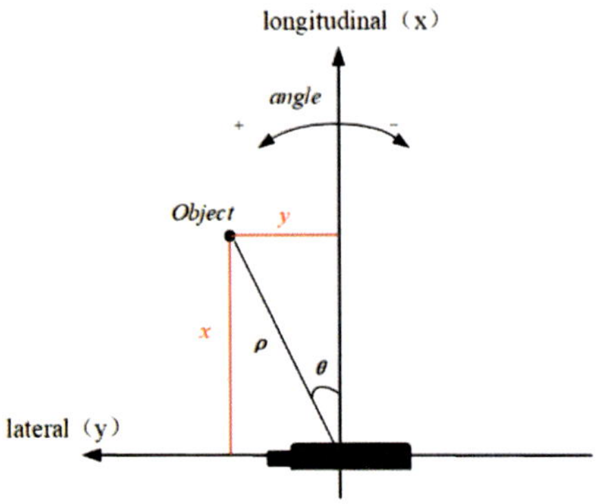

Fig. 2. Radar coordinate system

The conversion matrix is as follows:

$$\begin{bmatrix} x \\ y \end{bmatrix} = \begin{bmatrix} \rho \cos \theta \\ \rho \sin \theta \end{bmatrix} \tag{1}$$

(x, y) means the coordinate point of the object; θ means the object's horizontal course angle; ρ means the object's distance from the radar.

B. Establish radar coordinate system

Since the installation position of the millimeter-wave radar has a certain influence on the calculation of the train speed, and the object of this article is the train speed, to determine the train speed more accurately, the radar coordinates also need to be coordinated conversion, so we also need to establish the train as the center Cartesian coordinate system (see Fig. 3).

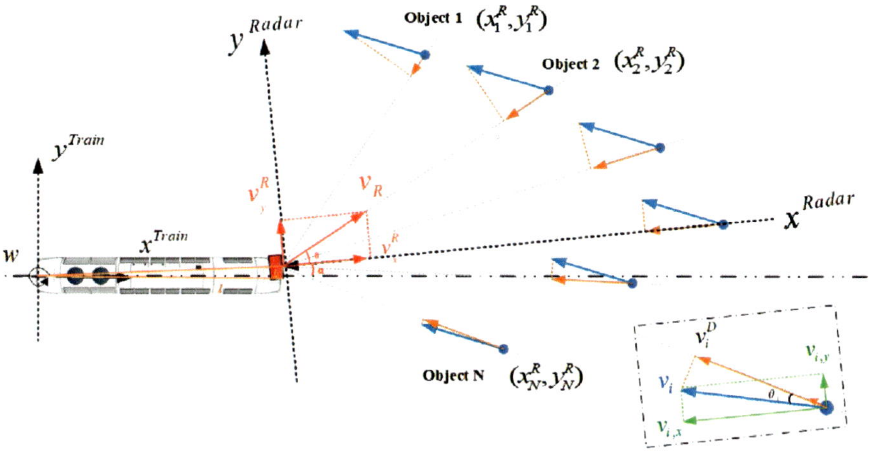

Fig. 3. The transformation from radar coordinate system to train motion coordinate system

This research uses the centroid of the millimeter-wave radar detector and the center of the train tail as the origin to establish a rectangular coordinate system. The train motion coordinate system is (x^T, y^T). The x^{Train}-axis is parallel to the longitudinal axis of the train and points to the front of the train; the y^{Train}-axis is parallel to the lateral axis of the train and points to the left of the train. The millimeter-wave radar is installed on the front of the train and is fixedly connected to the car body. The radar coordinate system is (x^R, y^R), the x^{Radar} axis is parallel to the radar beam, and the direction points to the front of the radar; the y^{Radar} axis is parallel to the lateral axis of the radar, and the direction points to the left of the radar. The process of converting coordinate points in the radar coordinate system to coordinate points in the train motion coordinate system is as follows:

$$\underbrace{\begin{bmatrix} x, T \\ y, T \end{bmatrix}}_{Train} = \begin{bmatrix} l + \frac{\cos\alpha}{\cos\theta} & 0 \\ 0 & b + \frac{\sin\alpha}{\sin\theta} \end{bmatrix} \underbrace{\begin{bmatrix} x, R \\ y, R \end{bmatrix}}_{Radar} \tag{2}$$

Finally, the object point in the train coordinate system can be written as:

$$\underbrace{\begin{bmatrix} x, T \\ y, T \end{bmatrix}}_{Train} = \begin{bmatrix} l + \rho \cdot \cos\alpha \tan\theta \\ b + \rho \cdot \sin\alpha \arctan\theta \end{bmatrix} \tag{3}$$

(x^T, y^T) means the coordinate point of the object; α means radar installation angle; (b, l) means the position of the radar relative to the train.

2.2 Preprocessing of Radar Measurement Data

According to the detection principle of millimeter-wave radar, although it can provide accurate position information and speed information of the object, there are still some problems. For example, if there is empty object information in the returned object, that is, the ID of the object is not detected, it will return to the system default value; There is invalid object information in the returned object, that is, the invalid interference object will be generated when the train bumps or speed changes and there are other disturbances. Therefore, before using radar data, certain methods and rules need to be used to eliminate invalid objects to ensure the accuracy of the train's perception of the environment. The data preprocessing process of millimeter-wave radar includes: radar data analysis, filtering out empty objects, filtering out duplicate objects, filtering out invalid objects, and extracting static objects.

(1) Filter out empty objects

The empty object meets the following conditions:

$$x_i^R = 0 \cap y_i^R = 0 \tag{4}$$

When both x_i^R and y_i^R are 0, it is judged to be an empty object, and the object that meets the above conditions will be filtered out.

(2) Filter out duplicate objects

If there is duplicate data in the returned object, the data that appears for the first time will be retained, and the rest will be deleted. The duplicate object meets the following conditions:

$$ID_i == ID_i \cap x_j^R == x_i^R \cap y_j^R == y_i^R \cap \ldots \cap v_{x,j}^R == v_{x,i}^R \cap v_{y,j}^R == v_{y,i}^R \tag{5}$$

(3) Filter out invalid objects

Millimeter-wave radar is often affected by noise, clutter, and multipath, resulting in many false objects, etc., which brings disadvantages to subsequent data processing. Therefore, it needs to be filtered out.

First, if the range, angle, and speed of the object are not within the range of the radar scanning capability, it is considered to be an unreasonable output point and removed; at the same time, the point cloud generated by factors such as noise and clutter is relatively isolated and scattered, so Use the positional relationship between the point clouds to perform filtering again. According to the position of the point cloud, search for a certain number of point clouds closest to the point cloud, calculate the average distance between the point and the neighboring point cloud, count the mean and standard deviation of all the average distances, and determine the distance threshold (mean + standard deviation) *4) If the average distance is greater than the distance threshold, these points are considered to be relatively isolated points and removed. Finally, due to the short existence of false objects, judging by the number of occurrences of the object, if the number of consecutive occurrences is less than 5, (the radar model used in this experiment is ARS-408 millimeter-wave radar, the sampling interval is about 72 ms) that is continuous If the appearance time is less than 0.36 s, the object is regarded as a false object (see Fig. 4).

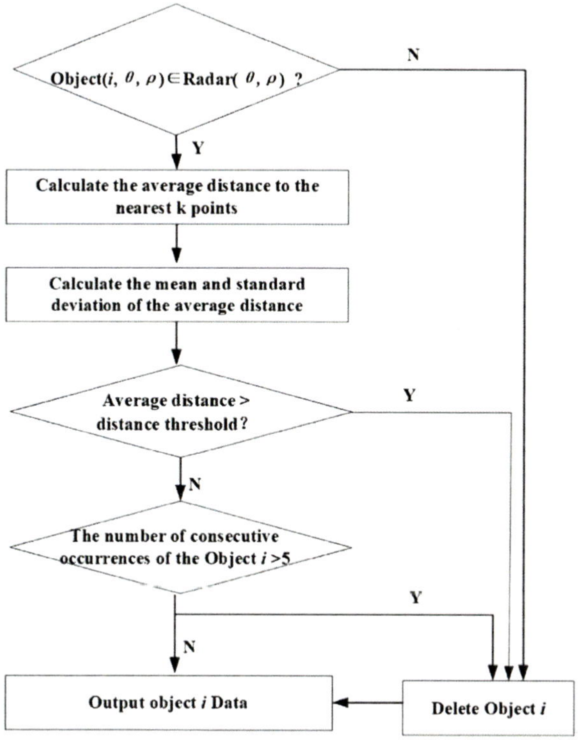

Fig. 4. Process of filtering out invalid objects

2.3 Static Objects Detection Algorithm

The preprocessed objects are classified into static objects and moving objects to extract a static object group. The recognition of static objects does not require precise positions, but only speed information. From the radar's point of view, all stationary objects move in opposite directions. The relative speed of the radar and the static object is exactly equal to the speed of the radar sensor. The Random Sample Consistency (RANSAC) algorithm used in this research was proposed by Fischler and Bors in 1981 [12], which is an iterative estimation method that helps to find the model supported by the largest number of given data points. Assuming that the correct object group all belongs to stationary objects, the error is a moving object that will produce a deviated velocity distribution. The algorithm process is as follows:

(1) Set the initial number of interior points of the optimal interior point S_0 (set 0), and determine the number of sampling N. n is the minimum amount of data required to calculate the model, w is the proportion of interior points, and p is the probability that the pair of matching points is interior points (p generally takes a value of 95%), then the number of samples is:

$$N = \frac{\log(1-p)}{\log(1-w^n)} \tag{6}$$

(2) For each object data $Object\_i(x_i^R, y_i^R, v_{x,i}^R, v_{y,i}^R, \theta_i), i \in \{1, 2, \cdots N\}$, randomly select the speed information of four objects $a_j(v_{x,j}^R, v_{y,j}^R), j \in \{1, 2, 3, 4\}$, calculate the homography matrix H, and record the parameter model corresponding to the matrix as M_0, and s is the scale parameter.

$$s\begin{bmatrix} v_x' \\ v_y' \\ 1 \end{bmatrix} = \begin{bmatrix} h_{11} & h_{12} & h_{13} \\ h_{21} & h_{22} & h_{23} \\ h_{31} & h_{32} & h_{33} \end{bmatrix} \begin{bmatrix} v_x^R \\ v_x^R \\ 1 \end{bmatrix} \tag{7}$$

(3) Calculate the Euclidean distance d_i between all the points to be matched $b_j(v_{x,j}^R, v_{y,j}^R), j \in \{1, 2, \cdots, N-4\}$ and the model. If $d_i < \varepsilon$ (the threshold is set to 3 in this research), add the inner point set S_0, that is, the correct matching point pair, otherwise it is the outer point. The threshold is used to identify abnormal values or moving objects and limit their errors within the threshold.
(4) If the elements in the current point set S_0 are greater than the optimal interior-point set S^*, and the current H matrix is the best estimation matrix, then update $S^* = S_0$.
(5) Jump to step (1), after N iterations or the number of interior points remain unchanged, and finally get the largest number of pointsS^*, then the model obtained by corresponding parameters at this time is the optimal match, point set S^* It is a static object group.

2.4 Train Speed Calculation

After getting the static object group, calculate the average speed of the static object group:

$$\underbrace{\begin{bmatrix} v_{r,\theta_1} \\ v_{r,\theta_2} \\ \cdots \\ v_{r,\theta_N} \end{bmatrix}}_{Radar} = \begin{bmatrix} \cos\theta_1 & \sin\theta_1 \\ \cos\theta_2 & \sin\theta_2 \\ \cdots & \cdots \\ \cos\theta_N & \sin\theta_N \end{bmatrix} \underbrace{\begin{bmatrix} v_x, R \\ v_y, R \end{bmatrix}}_{Radar} \tag{8}$$

$$\overline{v_{r,\theta}} = \frac{1}{N}\sum_{i=1}^{N} \underbrace{\begin{bmatrix} v_{r,\theta_1} \\ v_{r,\theta_2} \\ \cdots \\ v_{r,\theta_N} \end{bmatrix}}_{Radar} \tag{9}$$

v_{r,θ_N} means the speed of the static object point N in the radar coordinate system, $\overline{v_{r,\theta}}$ means the average speed of the static object group.

The radar speed value is the relative speed value of the radar and the static object group, and the direction is opposite to it.

$$v_{R,\theta} = -\overline{v_{r,\theta}} \tag{10}$$

$v_{R,\theta}$ means the radar speed.

The purpose of this research is to measure the speed of the train. Therefore, it is necessary to convert the sensor speed to the train speed through the coordinate transformation shown in Fig. 2.

$$\underbrace{\begin{bmatrix} v_x, T \\ v_y, T \end{bmatrix}}_{Train} = v_{R,\theta} \cdot \begin{bmatrix} \cos(\theta + \alpha) \\ \sin(\theta + \alpha) \end{bmatrix} \tag{11}$$

3 Experiments

3.1 Setup

The test train is equipped with an ARS-408 mm-wave radar installed in front of the train. The following Table 2 gives the parameters of this type of radar. The radar asynchronously sends detection signals every 72 ms, processes the reflected signals, and displays them in the form of objects after multiple steps, representing the historical trajectory and dimensions of the object. Among them, this article selects *ObjTimestamp, Obj_ID, Obj_DistLong, Obj_DistLa, Obj_VrelLong, Obj_VrelLat, Obj_DynProp, Obj_RCS* these 8 more important data indicators for research. The meanings are object timestamp, object longitudinal distance, object lateral distance, object longitudinal speed, object lateral speed, Object existence possibility, radar cross-sectional area.

The train is also equipped with an odometer (based on wheels and gyroscope) for speed measurement, with an operating time of 200 ms.

3.2 Results

A. **Data preprocessing algorithm verification**

By filtering out empty objects, duplicate objects, and invalid objects through data preprocessing, it can prepare for effective object tracking in the later stage. According to the previous algorithm, The following Table 3 gives the processing results.

Table 2. ARS-408 mm-wave radar system parameters

Parameters	Values
Detection distance	0.20…250 m long-range 0.20…70 m/100 m@0… ± 45° short-range 0.20…20 m@ ± 60°short-range
Cycle	About72ms
Azimuth	−9.0°… + 9.0°long range, −60°… + 60°shor-range
Azimuth beamwidth (3 dB)	2.2°long-range 4.4°@0° / 6.2°@ ± 45° / 17°@ ± 60°short-range
Speed resolution	0.37 km/h long range, 0.43 km/h short-range
Range resolution	long-range, 0.39 m short-range

Table 3. Data preprocessing results

Obj_ID	ObjTimestamp	Obj_DistLa	Obj_DistLong	Obj_VrelLong	Obj_VrelLat	Results
27	00:05:48:223	14.2	0.4	7.0	– 1.00	Valid
27	00:05:48:223	14.2	0.4	7.0	– 1.00	Duplicate
39	00:05:48:223	22.8	– 4.0	7.0	– 1.50	Duplicate
31	01:17:37:679	165.8	24.2	– 6.4	16.5	Invalid

Compare the data before and after the radar data processing to verify the effect of the tracking algorithm in filtering out the interfering objects. There are multiple invalid objects and repeated objects (Fig. 5). After valid object selection, the repeated objects and invalid objects are filtered out. The position information of the object is clearer (Fig. 6).

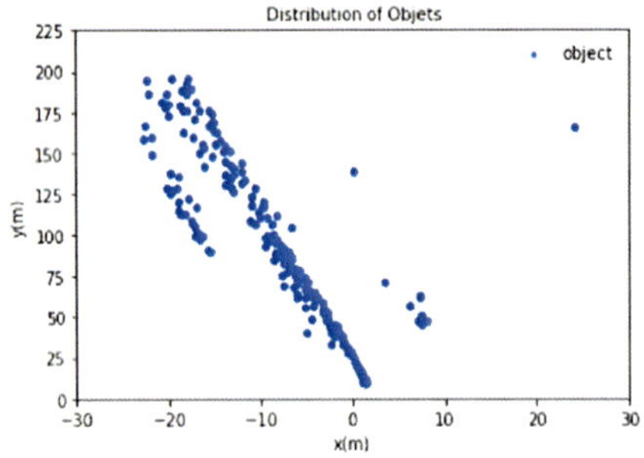

Fig. 5. Before data preprocessing

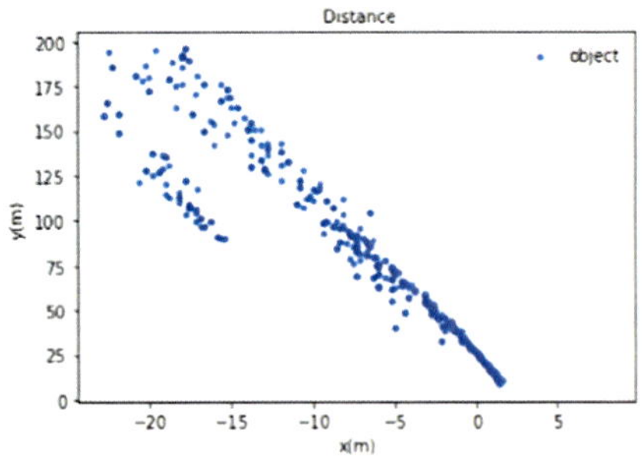

Fig. 6. After data preprocessing

B. Train speed calculation

The radar sensor calculates the train speed independently of the train. To analyze the speed measurement effect, compare it with the speed measurement method using the odometer. By measuring the entire sequence (about 250 s), you can see the comparison results (Fig. 7). The results show that even with a large number of moving objects or clutter, the system operation maintains continuous stability, has good technical performance and real-time performance, and can be used in actual projects.

To further analyze the effect of radar speed measurement, this research conducts residual analysis on radar measurement data and odometer measurement data and calculates the speed difference between radar and odometer (Fig. 8). The results show that the radar speed value is very close to the odometer speed value, and the residual fluctuates around 0. The reason for the large moment of error is that there are some factors in the actual situation (such as severe turbulence, human influence, etc.) that cause abnormal speed fluctuations, and the maximum error is about 1.5 m/s.

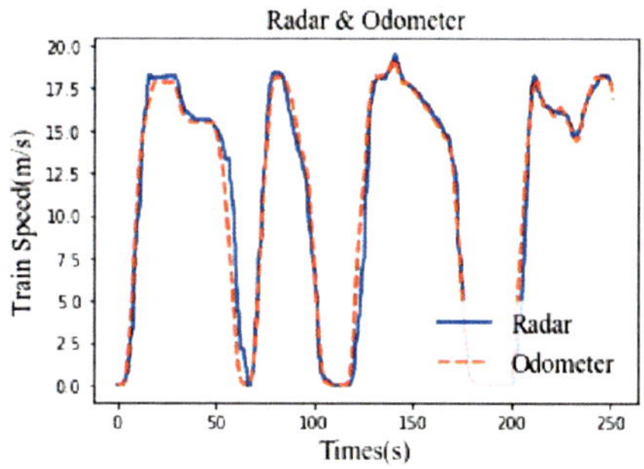

Fig. 7. Comparison of radar and odometer

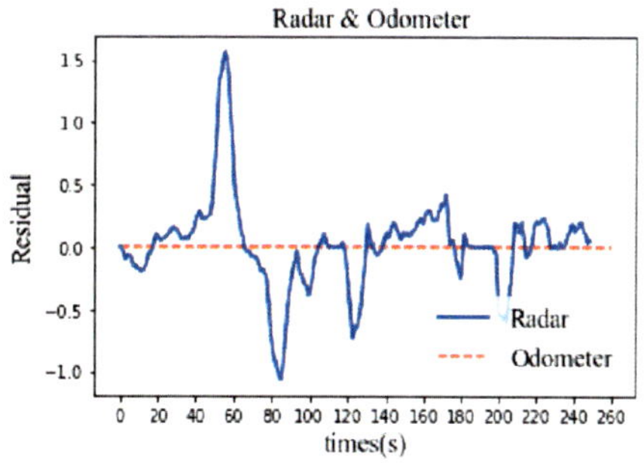

Fig. 8. Residual chart of radar and odometer

In addition, the speed error of radar and odometer is related to the running speed of the train. When the train accelerates or decelerates rapidly, the train measurement error is relatively large (see Fig. 9, 10); in the normal high-speed driving process, the speed measurement error is relatively small (Fig. 11).

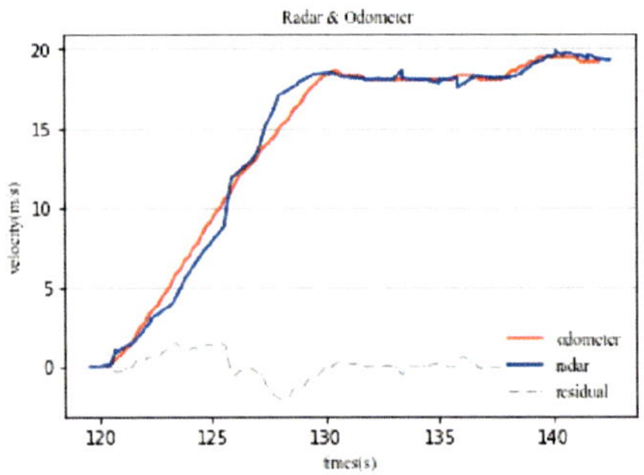

Fig. 9. Comparison of acceleration data

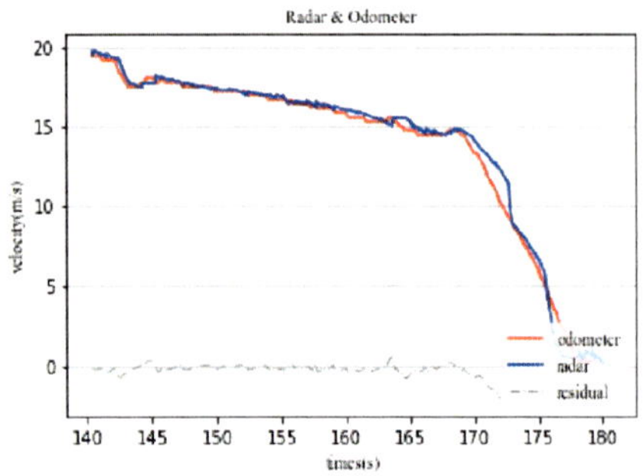

Fig. 10. Comparison of deceleration data

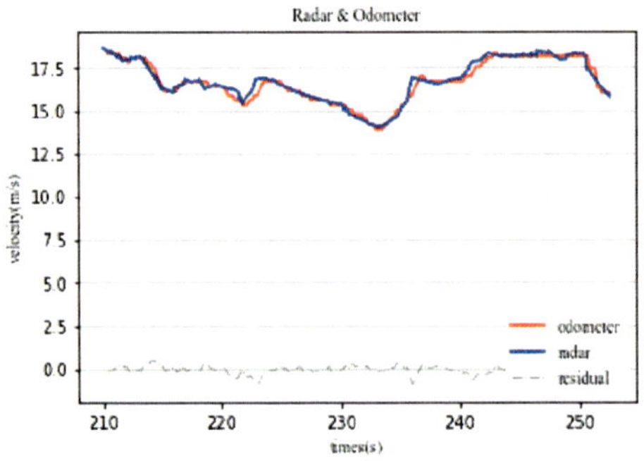

Fig. 11. Comparison of normal driving data

4 Conclusion

This research proposes a method of using millimeter-wave radar to achieve train speed measurement. This method takes advantage of the millimeter-wave radar to continuously get object motion information all-weather and realizes the calculation of the train's speed through the detection of static objects. Experiments show that the method of using millimeter-wave radar to measure train speed in the real track environment has better accuracy and stability, and is more sensitive than traditional odometer-based methods. The follow-up will further improve the system's object detection accuracy, enrich the capabilities of the train detection system, and play a greater role in the field of rail transit.

Acknowledgments. The project is supported by National key R&D projects (Number: 2022YFB4300501) and Shanghai Science and Technology Committee Project (Number: 23DZ2204900).

References

1. Zhou, D.T.: Research on multi-sensor information fusion based train location method. Jiaotong University, Beijing (2007)
2. Liu, J., Cai, B.G., Wang, J., et al.: Study on wheel diameter calibrastion method in integrated train positioning based on gray theory. J. China Railway Soc. **33**(5), 54–59 (2011)
3. Badino, H.: A Robust Approach for Ego-Motion Estimation Using a Mobile Stereo Platform. In: Jähne, B., Mester, R., Barth, E., Scharr, H. (eds.) IWCM 2004. LNCS, vol. 3417, pp. 198–208. Springer, Heidelberg (2007). https://doi.org/10.1007/978-3-540-69866-1_15
4. Luvizon, D.C., Nassu, B.T., Minetto, R.: A video-based system for vehicle speed measurement in urban roadways. IEEE Trans. Intell. Transp. Syst., 1–12 (2016)
5. Liu, J., Cai, B.G., Wang, J., et al.: Odometer-aided jamming detection method for satellite-based train positioning. J. China Railway Soc. **42**(10), 76–86 (2020)
6. Filip, A.: Efficient use of multi-constellation EGNOS for the European Train Control System (2016)
7. Liu, Y., Ning, B.: Discussion on distance measuring method of train velocimetry. Railway Signaling Commun. **12**, 3–5 (1997)
8. Zhang, X.M.: Ranging method and accuracy analysis based on vehicle laser radar. University of Electronic Science and Technology of China (2019)
9. Kellner, D., Barjenbruch, M., Klappstein, J., et al.: Instantaneous ego-motion estimation using Doppler radar. In: International IEEE Conference on Intelligent Transportation Systems (2013)
10. Kellner, D.: Instantaneous ego-motion estimation using multiple Doppler radar. In: IEEE International Conference on Robotics and Automation (2014)
11. Zhou, Y., Zhou, Q.L., Zheng, C.X., Zhang, Q.Y.: Rail transit speed measurement method and error analysis based on dual radar. Control Inf. Technol. (01), 30–34(2021)
12. Shi, X.: Application and development of millimeter-wave radars. Telecommun. Eng. **01**, 1–9 (2006)
13. Fischler, M.A., Bolles, R.C.: Random sample consensus: a paradigm for model fitting with applications to image analysis and automated cartography. Commun. ACM **24**(6), 381–395 (1981)

Study on Clustering Analysis Model of Traffic Congestion State

Kai Su and Yujie Sun(✉)

State Nuclear Electric Power Planning Design & Research Institute Co., Ltd, Beijing 100095, China
Kai.sukai@snpdri.com, Yujie.larris@126.com

Abstract. Considering the complexity of spatial and temporal changes of traffic data and the huge nature of traffic data, in order to find a sensitive analysis method for traffic congestion state changes, this paper conducts research on cluster analysis model. Firstly, the clustering analysis model is divided into four parts for discussion: clustering analysis algorithm, clustering analysis index selection, clustering number determination and index weight discussion. Then, the different methods contained in each part are recombined to obtain different cluster analysis models. Finally, three clustering analysis models are summarized: Model I, Model II and Model III. In the next stage, evaluation indicators will be selected for quantitative evaluation of the three models to verify the stability of the model method.

Keywords: Traffic Congestion State · Clustering Analysis Model · Quantitative Analysis Index · Model Selection

1 Introduction

Cluster analysis is widely used in various fields of engineering technology. In the research on traffic congestion and road network traffic state, people combine the cluster analysis method with the theory of traffic engineering, solving many practical problems.

On the basis of the classical traffic flow theory, Chen Feng et al. discussed the traffic state classification method combined with the congestion section and established the traffic state classification model based on the fuzzy C-means clustering. On this basis, Zhang Liangliang et al. applied traffic parameter weights to the fuzzy C-means clustering algorithm, and obtained the fuzzy C-means road traffic state division method based on parameter weights. Based on the clustering analysis of traffic flow characteristics, Kaneko et al. redefined the degree of congestion. Diamantopoulos et al. proposed a traffic congestion visualization method based on cluster analysis, which could identify the heavily congested areas in the road network. Based on this, they proposed a new traffic congestion prediction algorithm. Anbaroglu et al. proposed a detection method for occasional congestion in large urban networks, which was achieved by clustering analysis of road trip times. Yang Zuyuan et al. proposed a fast fuzzy C-means clustering (FFCM) algorithm for the identification of urban traffic congestion state, and applied the

X. Zeng et al. (Eds.): ITASC 2022, LNEE 1042, pp. 182–195, 2023.
https://doi.org/10.1007/978-981-99-2252-9_17

algorithm to the clustering analysis of urban traffic flow data. Zhao Yan et al. proposed an improved fuzzy C-means algorithm for the identification of urban road traffic state, and then proved the effectiveness of the method for the identification of road traffic state through simulation. Based on cluster analysis, Li et al. proposed an algorithm to predict background traffic volume, which effectively reduced the complexity of large-scale road network traffic simulation in the process of studying congestion characteristics. In the study of urban road network traffic comprehensive evaluation, Wang Yuanyuan et al. applied the principle of grey clustering analysis to solve the problems of subjectivity existing in multi-index evaluation. Thus, the subjective factors in the evaluation process are reduced. In terms of traffic congestion and traffic emissions, Li et al. conducted cluster analysis based on the temporal and spatial characteristics of traffic states in order to reduce the uncertainty of emission measurement. Jin Qiusi et al. chose speed as the intermediate variable to study the quantitative relationship between traffic operation index and motor vehicle fuel consumption emission intensity. The k-means cluster analysis method is used to study the velocity distribution characteristics of the traffic operation index, so as to analyze the uncertainty between the traffic index and the velocity distribution. Thus, the uncertainty of the relationship between fuel consumption emission intensity and traffic operation index is reduced.

In the process of establishing the cluster analysis model, every day is regarded as the sample point in the cluster analysis, and the corresponding indexes are the coordinate values of the points in each dimension of Euclidean space. The clustering analysis method is adopted to classify the dates so as to obtain different traffic congestion state.

In this paper, the clustering analysis model is divided into four parts for discussion: clustering analysis algorithm, clustering analysis index selection, clustering number determination and index weight. Each section contains several different methods, for example, clustering analysis algorithm contains hierarchical clustering method, the k-means clustering method algorithm, such as the selection of clustering analysis indexes according to the different standards can be divided in traditional traffic index curve geometry selection index, streamline index and embody curve change details directly. On this basis, this paper tries to find a sensitive analysis method for the change characteristics of traffic congestion state. In this paper, different methods contained in each part are recombined to obtain different cluster analysis models, which lays a foundation for subsequent comparison and selection to determine the optimal results.

2 Cluster Analysis Model

2.1 Selection of Clustering Analysis Algorithm for Traffic Congestion State

Cluster analysis algorithm is the core of traffic congestion state cluster analysis model. The general idea of clustering analysis algorithm is to classify the points close to each other step by step according to the different distances of each sample point and by using iterative and recursive methods. Sample points are regarded as points in n-dimensional space, and there are various distances between points, such as Euclidean distance, Block distance, Chebyshev distance, Minkowski distance, Mahalanobis distance, etc. In this paper, cluster analysis at Euclidean distance is discussed. Euclidean distance is calculated as follows.

$$d_{ij} = \left[\sum_{k=1}^{n}\left(x_{ik} - x_{jk}\right)^2\right]^{\frac{1}{2}} \tag{1}$$

where, d_{ij}—The Euclidean distance between points *i* and *j*,

n—Dimensions of Euclidean space,

x_{ik}, x_{jk}—The coordinates of points *i* and *j* in the *k*th dimension.

There are various types of clustering analysis algorithms, such as hierarchical clustering method, k-means clustering method, etc., which will be analyzed separately in this section.

Hierarchical Clustering Method. Hierarchical clustering method is also called system clustering method. Its basic idea is to combine the two closest sample points in the space, take the middle point of the two points as a new point, and calculate the distance between the point and other points. Repeat the above process until the number of remaining points meets the clustering number requirement. The specific operation steps are as follows.

Each sample point among m sample points is regarded as a point distributed in n-dimensional space, and n is the number of indexes of cluster analysis;

Find the distance between pairs of m points, and you get m(m-1)/2 distances;

Find the minimum distance value, combine the corresponding two points, take the midpoint of the two-point line as the new point, and return step 2), where m is the number of remaining points; If the remaining number of points happens to be the required cluster number N, stop merging and enter step 4);

Each final point is traced back, and the initial sample points it contains are grouped together.

In practical applications, it is usually not necessary to combine all the points together, but to divide them into several categories as needed. Draw a vertical line in the tree diagram. The number of horizontal lines intersecting the vertical line represents the clustering number, and the left sample point connected by each intersecting horizontal line is the point contained in the category. If it is necessary to divide all points into four categories, draw a red line that intersects four horizontal lines. The point to the left of each horizontal line that intersects falls into one category. If you want to divide it into 5 categories, draw a blue perpendicular.

It can be seen from the above process that in the process of operation, the hierarchical clustering method only differs in the degree of point merging for different clustering numbers, and its calculation method is completely the same. This shows that for the hierarchical clustering method, the results of different clustering numbers can be determined only by one calculation. In addition, its algorithm is simple and clear, and the calculation process is intuitive, which are the advantages of hierarchical clustering method over other clustering algorithms.

K-means Clustering Method. The basic idea of K-means clustering method is to select a number of initial points as the initial clustering center, and put the rest points into each center according to the principle of nearest distance, so as to obtain the first iteration result. Then, the center of gravity of each category is taken as the center of the next iteration to carry out repeated iteration, and the final result gradually converges and approaches the optimal solution.The specific operation steps are as follows:

In n-dimensional Euclidean space, N points are randomly selected as initial points according to the determined cluster number N;

Take these N points as the center, calculate the Euclidean distance from the rest points to these N points, and classify each point and its nearest center point into the same category;

Take the center of gravity of each category obtained in step 2) as the new center point and return to step 2); After iteration, the change of center of gravity is less than the set maximum threshold. Then, the iteration is stopped and step4) is entered.

At the end of clustering, the result classified before the last iteration is the final result.

The common characteristic of the above two clustering analysis algorithms is that the calculation principle is relatively simple, and it is easy to be implemented in statistical software such as MATLAB and SPSS. These two methods are widely used in practice. This paper use these two methods to build a clustering analysis model of traffic congestion state.

Cluster Analysis Index of Traffic Congestion State. Cluster analysis index is an important parameter in the process of establishing cluster analysis model and is the basis of cluster analysis algorithm. If each date is regarded as a point in Euclidean space, then the selected cluster analysis index determines the position of this point in space and the spatial relationship between each point. Where, the number of selected indexes corresponds to the dimension of the space, and the value of the indexes corresponds to the coordinates of each dimension. If the indexes are properly selected, the characteristics of each date and their interrelationships can be clearly and completely displayed, and the cluster analysis based on this can also obtain scientific and reasonable results.

Based on the data of Beijing in September 2018, Wen et al. proposed a set of index system based on the geometric shape of traffic index curve. In order to carry on the analysis comparison, this paper will propose another two kinds of new index system on this basis.

Clustering Index Based on Geometric Shape of Traffic Index Curve. Wen et al. found that all the daily traffic index curves had certain commonness, that is, during the morning peak (7:00–9:00, holiday delay) and evening peak (1700–18:00), there were two peaks on the curve, and the other time periods were flat peaks. Therefore, the geometric shape of the curve might be related to the indexes shown in Table 1.

The indexes of $X_1 \sim X_8$ are called "traditional indexes". The traditional index starts from the geometric shape of the traffic index curve, takes into account the common characteristics of the traffic index curve (with two peaks in the morning and in the evening), makes an in-depth analysis of the characteristics of the two peaks, and takes into account statistical indexes such as maximum value, mean value and variance. In addition, X_7 and X_8 indexes also include the feature of considering the length of congestion throughout the day (the number of severe congestion and moderate congestion points throughout the day).

Table 1. Traditional Indexes

Index	Practical Implications	Curve Implications
X_1	Maximum of the morning peak index	Highest point of the first crest
X_2	Maximum of the evening peak index	Highest point of the second crest
X_3	Average index during the morning peak hour	Position of the first crest
X_4	Average index during the evening peak hour	Position of the second crest
X_5	Variance of the indices during the morning peak	Measure the width of the first crest
X_6	Variance of the indices during the evening peak	Measure the width of the second crest
X_7	Number of serious congestion points	Measure the local position characteristics of the curve
X_8	Number of moderately congested points	Measure local position characteristics

However, the traditional indexes also have the problems of too many indexes and redundancy. Zhang Weiqun believes that in addition to the principle of comprehensiveness, attention should also be paid to the principle of refinement in the construction of the index system. In other words, the index system should not only reflect the overall information, but also reduce the number of indexes. Multiple indexes, such as average value and maximum value, jointly describe the characteristics of the peak value, which may also lead to the phenomenon of relatively large correlation between indexes. For the selection of indexes, this will obviously cause too much repetition and produce useless indexes.

Taking the data in September 2018 as an example, the correlation between traditional indexes is analyzed.

After preprocessing the index data, the index values of each date are obtained, as shown in Table 2.

Pearson correlation coefficient was used for correlation analysis of the above indexes $X_1 \sim X_8$, and the results were shown in Table 3.

Among the traditional indexes, the correlation coefficients of the maximum morning peak index X_1, the average morning peak X_3, the maximum evening peak index X_2 and the average evening peak index X_4 are all over 0.980, showing a high correlation. This is obviously not appropriate for cluster analysis. Therefore, the above indexes need to be streamlined.

Simplified Indexes Based on the Geometric Shape of a Traffic Index Curve. In order to reduce the number of indexes and the influence of auto-correlation between indexes, this paper proposes a method of simplifying indexes (Table 4) based on geometric features of traffic index curves.

Each index is explained in detail below.

Table 2. Illustration of Traditional Indexes

Date	X_1	X_2	X_3	X_4	X_5	X_6	X_7	X_8
9/1	1.9	1.6	4.3	3.5	0.2	0.6	0	0
9/2	8.5	7.6	4.8	4.0	1.1	0.6	4	2
9/3	8.2	7.1	7.4	6.3	1.3	1.1	3	8
9/4	8.2	6.9	5.5	4.8	1.2	0.6	2	4
…	…	…	…	…	…	…	…	…
9/30	3.4	2.8	3.3	3.1	0.5	0.3	0	0

Table 3. Results of the Correlation Analysis for The Traditional Indexes

Pearson Correlation Coefficient	X_1	X_2	X_3	X_4	X_5	X_6	X_7	X_8
X_1	1	0.631	0.994	0.591	0.940	0.339	0.498	0.507
X_2	0.631	1	0.588	0.986	0.700	0.421	0.577	0.791
X_3	0.994	0.588	1	0.549	0.904	0.312	0.507	0.470
X_4	0.591	0.986	0.549	1	0.660	0.274	0.652	0.803
X_5	0.940	0.700	0.904	0.660	1	0.376	0.443	0.566
X_6	0.339	0.421	0.312	0.274	0.376	1	−0.266	0.222
X_7	0.498	0.577	0.507	0.652	0.443	−0.266	1	0.621
X_8	0.507	0.791	0.470	0.803	0.566	0.222	0.621	1

Total Traffic Congestion. In order to prevent interference caused by low sample size at night, traffic index data from 6:00 to 23:45 are taken and simply added to obtain the total traffic congestion X9. This is an overall measure of traffic congestion throughout the day.

Comparison between Morning and Afternoon. Take the sum of the exponentials in the afternoon (12:00 – 23:45) and the sum of the exponentials in the morning (6:00 – 11:45) and divide them to get the ratio X_{10} in the morning and afternoon. This value reflects the relative relationship between the congestion in the moring and the afternoon.

The Prominence of the Congestion Peak. Otsu grayscale segmentation method was used to segment the traffic index from 6:00 to 23:45. Otsu segmentation method is usually used in image processing. Its basic idea is to minimize the intra-class variance and maximize the inter-category variance. For a black and white image, the gray value of each pixel is different. Otsu method is used to obtain a gray threshold, so that the gray value of all points greater than this threshold is changed to 255 (black) and that of all pixels less than this threshold is changed to 0 (white), and a simplified binary image is obtained. The operation process is as follows:

Table 4. Simplified Indexes

Index	Practical Implications	Calculation Method
X_9	Total traffic congestion	Add traffic index data from 6:00 to 23:45
X_{10}	Comparison between morning and afternoon	The sum of the traffic index from 12:00 to 23:45 divided by the sum of the traffic index from 6:00 to 11:45
X_{11}	Prominence of the congestion peak	Otsu gray level segmentation method is used to perform binary segmentation of traffic index data from 6:00 to 23:45, and the proportion of "1" is obtained
X_{12}	Maximum change rate	The traffic index values from 6:00 to 23:45 are smoothed (the first and last values are unchanged, and the median value is the average value of the three Numbers: itself, front and back). Carry out difference processing, take the modulus, and take the average of the top 10 change rates

First, calculate the average gray level of the whole image (Eq. (2)).

$$u = \sum_i \frac{i \cdot n(i)}{P \cdot Q} \tag{2}$$

where, u—the average grayscale of the entire image;

n(i)—the number of points whose grayscale is i;

P·Q—The number of pixels in an image;

Set t as the threshold value, (i>t).

$$\omega_1 = \frac{W_1}{M \cdot N} \tag{3}$$

$$\mu_1 = \sum_i \frac{i \cdot n(i)}{W_1}, \tag{4}$$

where,ω_1—The percentage of points whose grayscale is greater than the threshold t;W_1—The number of points whose grayscale is greater than the threshold t;

μ_1—The average gray of all points where the gray level is greater than gray threshold t.

In the same way, the proportion ω_2 of points whose gray level is less than the threshold value and the average gray level μ_2 are calculated.

$$G = \omega_1 \cdot (\mu_1 - \mu)^2 + \omega_2 \cdot (\mu_2 - \mu)^2 = \omega_1 \omega_2 (\mu_1 - \mu_2)^2 \tag{5}$$

In Eq. (5), under the circumstance that G value is the largest, the corresponding t is the threshold X_{11}, which reflects the proportion of the congestion time in the whole day.

Maximum Change Rate. First, in order to reduce the impact of abnormal points on the results, the curve was smoothed. Specific method: the index values of 0:00 and 23:45 remain unchanged, and the values of the rest moments are taken as the average of the values of the preceding, the posterior and the self, so as to obtain the new curve. Then, find the absolute value of the difference between the exponents at the next two moments. Finally, average the maximum 10 differences to get the maximum rate of change X_{12} for a given day, which reflects the change rate of congestion during peak periods.

The above simplified index not only considers the characteristics of morning and evening peak, but also takes into account the index of non-peak period (such as the comparison of index X10 in the morning and afternoon and the peak prominence of index X_{11}). The characteristics of traffic operation reflected by the indexes are more comprehensive and have been greatly improved compared with the traditional indexes. Then, the data of September 2018 is also taken as an example to analyze the correlation between streamlined indexes (Table 5 and Table 6).

Table 5. Illustration of Simplified Indexes

Date	X_9	X_{10}	X_{11}	X_{12}
9/1	178.1	2.3	0.1	0.3
9/2	201.2	1.0	0.1	1.1
9/3	250.2	1.3	0.1	1.1
9/4	225.7	1.3	0.0	1.0
…	…	…	…	…
9/30	164.6	1.9	0.1	0.4

Table 6. Results of Correlation Analysis for the Simplified Indexes

Pearson Correlation Coefficient	X_9	X_{10}	X_{11}	X_{12}
X_9	1	−0.018	0.615	0.661
X_{10}	-0.018	1	0.197	-0.656
X_{11}	0.615	0.197	1	0.160
X_{12}	0.661	-0.656	0.160	1

It can be found that the absolute value of the correlation coefficient between all the indexes drops below 0.7, indicating a significant reduction in the correlation between the indexes.

Table 7. Comparison of the Three Clustering Models of Traffic Congestion State

Name	Cluster Analysis Algorithm	Clustering Index	Cluster Category Determination Method	Any Weight or Not
Model I	Hierarchical Clustering Method	Traditional Index	Manually Determined	No Weight
Model II	K-means Clustering Method	Simplify Index	Use the Highest Silhouette measure to determine	No Weight
Model III	K-means Clustering Method	Direct Index	Use the Highest Silhouette Measure to Determine	Use Coefficient of Variation to Weight

In the process of cluster analysis, sometimes the order of magnitude of the index value is inconsistent (as the traditional index and the simplified index introduced earlier). In order to prevent the interference of the index with a large order of magnitude to the results, the data should be standardized.

$$x'_{ik} = \frac{x_{ik} - \overline{x_k}}{S_k}; i = 1, 2, \ldots, n;\ k = 1, 2, \ldots, m \tag{6}$$

where, x' ik—The kth index of normalized i;x_{ik}—The original value of the kth index of i;$\overline{x_k}$—The average of all the points with the kth index;

$$\overline{x_k} = \frac{1}{n}\sum_{i=1}^{n} x_{ik} \tag{7}$$

S_k—The standard deviation of the kth index at all points;

$$S_k = \sqrt{\frac{1}{n-1}\sum_{i=1}^{n}(x_{ik} - \overline{x_k})^2} \tag{8}$$

Standardized data have the same dimensions and orders of magnitude.

An Index Based on the Continuous Time-Varying Difference of Traffic Index The traditional and simplified indexes are analyzed mainly from the geometric shape of the traffic index curve. In order to describe the slight changes of the traffic index curve more accurately, this paper tries to propose a new index system from the traffic index itself.

If you every moment of every day of traffic index as different indexes (in order to rule out floating car at night, the error of the small sample size used at about 6:00 – 21:00, 15 h of data), so, every day can be seen as one who has a 60 index of sample points, which is a 60 points in the Euclidean space, using the cluster analysis algorithm to cluster these points to get the final result. Since it directly uses traffic index as clustering index, it is called "direct index".

Categories of Traffic Congestion State. The determination of traffic congestion state categories, mainly solves the problem of dividing existing dates into several categories. Ideally, without considering holidays, rainfall days and traffic control days for special events (such as major meetings, odd-even license plate days caused by smog), the distribution of traffic congestion state should take a cycle of one week and show regular changes. This is due to the commuting nature of people's travel during rush hours. Assuming that five-day work day and two-day weekend have different raffic congestion State respectively, and considering the influence of holidays, bad weather and special events, it is reasonable to set the upper limit of all date classification Numbers to ten categories in the clustering process. In addition, too few classification numbers obviously do not accord with the actual situation. Considering the most intuitive classification of date, the lower limit of the number of classification is temporarily set as six categories.

Artificial Determine. In previous studies, the determination of cluster number is not strictly calculated, but manually determined. After determining the cluster number manually, the accuracy of the results remains to be further verified.

Silhouette Measure Calculation. Silhouette measure (Sil(a)) is a measure that measures the tightness and separation of categories within a sample point clustering result and can be used to determine the optimal number of categories. Sil(a) $\in$ [−1,1], the greater the value, the better the clustering quality. The calculation method is as follows.

$$Sil(a) = \frac{a_a - b_a}{\max\{a_a, b_a\}} \tag{9}$$

$$a_a = \min_{K' \neq K(a)} \left\{ \text{mean}_{A_c \in K'} |A_a A_c| \right\} \tag{10}$$

$$b_a = \text{mean}_{A_b \in K(a)} |A_a A_b| \tag{11}$$

where, *Sil(a)*—Aa's Silhouette measure;

a_a—Minimum average distance between Aa and samples in other categories;

b_a—Average distance between Aa and other samples in its category;

A_a—Sample;

A_b—A sample belonging to the same category as Aa;

A_c—A sample belonging to the different category as Aa;

K(*a*)—The category Aa belongs;

K'—Different categories from K(a).

Above calculation, a Silhouette measure of a sample point can be obtained. Take the average value of Silhouette measures at all points and then get the average value of Silhouette measures $\overline{S_N}$.

$$\overline{S_N} = \frac{1}{m} \sum_{a=1}^{m} Sil(a) \tag{12}$$

where,*m*—Number of sample points;

N—Category number.

For $\overline{S_N}$, when its value is the largest, the corresponding cluster number N is the optimal cluster category number.

Clustering Index Weight of Traffic Congestion State. In cluster analysis, the "importance degree" of all indexes is different, and different indexes should be treated differently. The weight problem discussed in this paper refers to the different coefficients multiplied by the corresponding value of each index according to its different importance in the clustering analysis of traffic congestion state. The weight of clustering index has been involved in some existing studies, but in the clustering analysis of traffic congestion state, there has been no relevant research.ptimal cluster category number.

Significance of Clustering Index Weight of Traffic Congestion State. In the clustering analysis, the weight of different indexes represents that the coordinate axes of each dimension in the N-dimensional space are stretched by different multiples, and the result will lead to the change of the distance between points in the space, thus affecting the result of the clustering analysis.

As shown in Fig. 1 (a), when there is no weight (or the weight is the same) for each coordinate axis, the distance from point D to point B and C is $\sqrt{8}$ and 3 respectively, and the nearest point from D is B. In Fig. 1 (b), after the weight of 2 is assigned to the X-axis, the X-axis is stretched twice as much, and its unit length is twice as much as the unit length of the Y-axis. It can be clearly observed that the relative position relationship between points has changed. The distance between D and B has become $\sqrt{20}$, and the distance between D and C is still 3, so obviously, the closest point to D has become C. Such a change in clustering analysis will have a significant impact on the results.

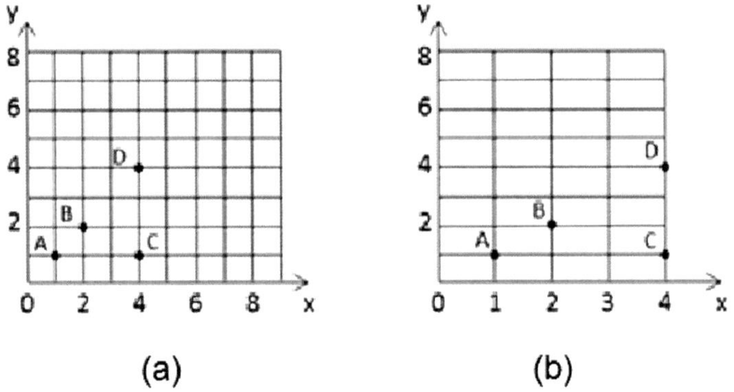

Fig. 1. Significance of Index Weight in Cluster Analysis

Clustering Index Weight Selection of Traffic Congestion State. The principle to select the weight of clustering index is that the curve of traffic congestion state should have a large degree of differentiation. For the index with big difference between different dates, a larger weight should be given to make it more fully differentiated. For the indexes with less difference, the weights should be given less to weaken their effect. Therefore, the coefficient of variation reflecting the degree of dispersion between data is considered as the weight in this paper.

The calculation method of the variation coefficient corresponding to the index is as follows.

$$CV = \frac{\sigma(x)}{E(x)} \tag{13}$$

$$E(x) = \frac{1}{m}\sum_{i=1}^{m} x_i \tag{14}$$

$$\sigma(x) = \sqrt{\frac{1}{m}\sum_{i=1}^{m} [x_i - E(x)]^2} \tag{15}$$

where, CV—Coefficient of variation corresponding to the index;

$\sigma(x)$—population standard deviation of all date index values;

$E(x)$—Average of the index values for all dates;

x_i—Index value of the ith date;

m—Dates Number to be clustered.

It can be seen from the above formulas (13) – (15) that the greater the degree of dispersion (difference) between the date index values, the greater the coefficient of variation, and correspondingly, the index will also be given a greater weight. On the contrary, the index with a smaller degree of value dispersion is given a smaller weight. It is worth pointing out that if all the dates have the same value for a certain index, then since its standard deviation (x) is 0, the corresponding coefficient of variation will also be 0. In other words, the weight of this index is 0, so it has no effect in clustering analysis and will be abandoned. That's the extreme case of weights.

Based on the above discussion on weight, it can be seen that, using coefficient of variation as weight, different weights can be assigned to different indexes according to the dispersion degree of different date index values. In this way, the importance of different indexes can be reflected, and the result of cluster analysis will be improved compared with the time when weights are not used.

3 Cluster Analysis Model Establish of Traffic Congestion State

After the discussion of different methods in four parts of clustering analysis, clustering analysis algorithm, clustering index selection, clustering category determination and weight determination, the following three clustering analysis models of traffic congestion state are determined in this paper (Table 7).

Model Based on Hierarchical Clustering and Geometric Analysis of Exponential Curves (Model I). In this model, hierarchical clustering method is used in clustering algorithm, and eight traditional indexes based on geometric shapes of traffic index curves are used in index selection. In addition, the model does not consider the determination of clustering category and weight analysis, but directly determines the clustering category artificially without considering the weight. The model is characterized by simple and direct methods, but the subjectivity is too large in the process of model construction.

Model Based on K-Means Clustering Method and Geometric Analysis of Exponential Curve (Model II). Based on the clustering analysis algorithm, this model selects the K-means clustering method with iteration as the basic feature. The selection of clustering index is still based on the geometric shape of traffic index curve. However, considering the potential redundancy and excessive correlation between traditional indexes in model I, the number of indexes was reduced to four. In addition, compared with model I, the model determines the clustering category using the highest Silhouette measure, but does not consider the weight between the indexes.

Model Based on K-Means Clustering Method and Characteristics of Exponential Continuous Time Variation (Model III). In terms of the selection of clustering indexes, this model considers the time-varying condition of traffic index throughout the day, which is more comprehensive than the indexes based on geometric shapes of traffic index in model I and II. At the same time, the model also considers the weight of each index and gives full play to the function of each index.

Among the three clustering analysis models, model I is the same as the model used in Wen's research. The other two models are new models proposed after the analysis and discussion of the four parts mentioned above. In the following research direction, indexes will be selected to evaluate these three clustering analysis models, and the best one will be selected as the final clustering analysis model of traffic congestion state.

4 Conclusion

For traffic congestion, different types of traffic (such as expressways, urban mainlines, and even rural roads) may have different characteristics. Based on the analysis of traffic congestion index and traffic index curve, this paper explores the cluster analysis model of traffic congestion state in four parts, cluster analysis algorithm, cluster index selection, cluster number determination and index weight. After detailed analysis of the different methods contained in each section, three clustering analysis models are finally concluded: Model I, Model II, and Model III. These three clustering analysis models lay a foundation for the following model comparison and clustering analysis of raffic congestion State.

Acknowledgments. The project is supported by Scientific research project of State Nuclear Electric Power Planning Design & Research Institute Co., Ltd (Number 100-KY2019-HYK-Z01).

References

1. Shao, C.F., Weili, Y., Jia, B.: Traffic Flow Theory, 1st edn. Electronic Industry Press, Beijing (2012)
2. Chen, F., Jia, Y., Niu, Z.: Study on traffic state partition method based on fuzzy c-means clustering. In: Proceedings of 2010 Asia-Pacific Conference on Information Theory (APCIT 2010) (2010)
3. Zhang, L., Jia Y., Niu, Z., Liao, C.: Traffic state classification based on parameter weighting and clustering method. J. Transp. Syst. Eng. Inf. Technol. (06),147–151 (2014)
4. Kaneko, Y., Ohe, I., Kawashima, H., Hirano, T.: The judgement of the traffic condition by using the cluster analysis. In: Presented at Pacific Rim TransTech Conference. Vehicle Navigation and Information Systems Conference Proceedings. 6th International VNIS. A Ride into the Future, Seattle, (2004)

5. Diamantopoulos, T., Kehagias, D., König, F., Tzovaras, D.: Use of density-based cluster analysis and categoryification techniques for traffic congestion prediction and visualisation. In: Presented at Transport Research Arena 2014, Paris, (2014)
6. Anbaroglu, B., Heydecker, B., Cheng, T.: Spatio-temporal clustering for non-recurrent traffic congestion detection on urban road networks. Transp. Res. Part C **48**(11), 47–65 (2014)
7. Yang, Z., Huang, X., Du, C., Tang, M.: Study of urban traffic congestion judgment based on FFCM clustering. Appl. Res. Comput. **25**(9), 2768–2770 (2008)
8. Fengbo, Z.H.A.O., Min, S.U.N.: Traffic state recognition method based on fuzzy cluster analysis. Microcomput. Appl. **22**(02), 9–13 (2006)
9. Li, T., Liu, J.: Cluster-based spatiotemporal background traffic generation for network simulation, Miami, United States. ACM Trans. Model. Comput. Simul. **25**(4), 1–25 (2014)
10. Wang, Y., Lu H., Qin, X.: Urban road network comprehensive evaluation based on gray cluster analysis. J. Highway Transp. Res. Dev. 22(08), 118–121 (2005)
11. Li, M., Yu, L., Song, G., Xie, L.: Traffic state clustering approach to lower the estimation uncertainty of greenhouse gas emissions based on traffic spatial and temporal characteristics. In: Transportation Research Board Annual Meeting (2016)
12. Wen, H., Sun, J., Zhang, X.: Study on traffic congestion modes of large city in china taking Beijing as an example. In: The 9th International Conference on Traffic & Transportation Studies (ICTTS'2014), Shaoxing (2014)
13. Weiqun, Z.: Research on construction of index system and method of optimal benign evaluation. Stat. Inf. Forum. **21**(06), 36–39 (2006)

Association Rules Mining for Railway Accident Causes Based on Improved HFACS

Xiaoqing Zeng[1](✉), Haixiang Lin[1,2], Ran Lu[2], and Gu Min[3]

[1] Key Laboratory of Road and Traffic Engineering of the Ministry of Education, Tongji University, Shanghai 201804, China
zengxq@tongji.edu.cn

[2] School of Automation and Electrical Engineering, Lanzhou Jiaotong University, Lanzhou 730070, China

[3] Shanghai Municipal Engineering Design Institute(Group) Co., Ltd., Shanghai 200092, China

Abstract. Aiming at the problem of full quantitative analysis of railway accident causes, an improved HFACS model is proposed. Firstly, based on the original HFACS model, an improved HFACS model for railway industry was constructed Secondly, based on the improved model and association rule algorithm, the causative factors of 504 railway accidents from 2008 to 2009 collected by a railway bureau were comprehensively quantified, the association rule base of accident causative factors was mined, and the data of accident causative factors were visualized. The results show that the four main causes of railway accidents include irregularities, inadequate safety inspection, inadequate awareness of safety responsibility and inadequate education of safety responsibility. Based on the improved HFACS causative correlation analysis method, the importance of technical factors can be enhanced. Finally, the author puts forward some solutions to the lack of safety responsibility consciousness of the key accident factors.

Keywords: Railway Accident · Improved HFACS Model · Causation Factors · Apriori Algorithm

The advantage of association rules lies in the visualization of association rules after mining the causative association rules of railway accidents. The close relationship among the causative factors of railway accidents can be shown more intuitively through images, and then further solutions can be proposed. In order to make the image easy to analyze, the first 20 strong association rules are visualized in this paper. Figure 2 shows a visualization of the top 20 strong association rules sorted by support.

1 Introduction

At present, with the rapid development of China's railway industry, more and more attention is paid to 'the occurrence of railway accidents. The occurrence of railway accidents has caused serious economic losses, casualties and so on, so it is urgent to analyze the railway accidents and draw lessons from them. The frequent occurrence of railway accidents seriously affects the competitiveness of China in the international

X. Zeng et al. (Eds.): ITASC 2022, LNEE 1042, pp. 196–204, 2023.
https://doi.org/10.1007/978-981-99-2252-9_18

railway. Therefore, the prevention of railway accidents is imperative. HFACS (Human Factors Analysis and Classification System), proposed by Wiegmann et al., is an accident cause model for safety accidents in aviation field. The model analyzes the failure factors at the four levels of unsafe behavior, preconditions of unsafe behavior, supervision of unsafe behavior and organizational influence in detail. Therefore, this model is not only applied in the aviation field, but also in the navigation industry, medical industry, coal mining industry and railway industry.

In the field of railway industry, Australian scholar Lisa Punzet et al. used HFACS applicable to the railway industry to analyze the investigation report (N = 35) of SPADS by the Australian Railway Fredding Organization to check the human factors involved in the incident and determine the future trend. British scholar Madigan Ruth et al. used HFACS model to understand the relationship between active factors and potential factors of railway safety accidents, as well as specific causal paths. In China, Zhan Qingjian applied the HFACS model to the analysis of railway accidents in China and identified the main causes of accidents. Chen Ruiwei uses HFACS model to conduct qualitative analysis on the hazard sources of high-speed railway traffic dispatching system from three perspectives of "man-machine-loop". Because the traditional HFACS model is not fully applicable to the railway industry, and the research on the HFACS model in the railway field is not in-depth at home and abroad. Therefore, an improved HFACS model suitable for railway field is proposed.

2 Improvement of HFACS Model

Because the railway accident is caused by the interaction of human, machine, environment, management and other factors. However, the traditional HFACS mainly analyzes the causes of accidents from the human aspect, without taking into account other important accident drivers, such as technical aspects: equipment ineffectiveness, equipment design defects and so on, which have a great impact on the occurrence of railway accidents. As a result, the traditional HFACS model is designed for accidents in the aviation field, but is not applicable to the railway field in some aspects. Therefore, it is necessary to improve the traditional HFACS.

After the statistics and analysis of 504 railway accidents, the traditional HFACS was improved. The improved HFACS is used to analyze the leading factors of railway accidents from four levels: unsafe behavior, premise of unsafe behavior, dereliction of duty by relevant railway departments, and organizational influence. The premise of unsafe behavior is again divided into human factors, technical factors, environmental factors. Raising the technical environment of the original environmental factors to a level reflects the importance of technical factors. The dereliction of duty by relevant railway departments has replaced unsafe supervision, which further reflects the intensity of railway management and control. In terms of organizational influence, it is further divided into three aspects: professional training, working conditions, and information communication. Figure 1 shows the improved HFACS model.

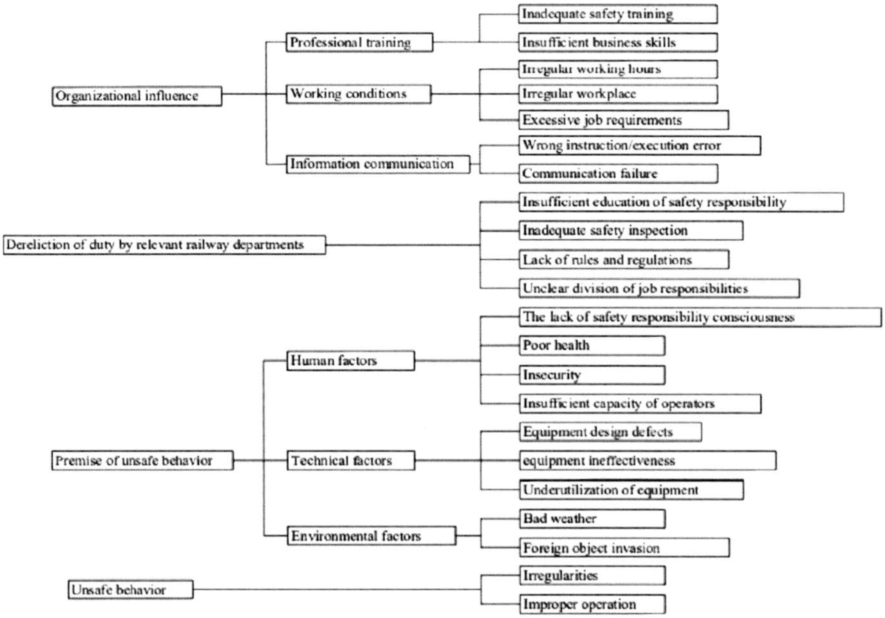

Fig. 1. Improved HFACS model

3 Correlation Analysis of Causative Factors of Railway Accidents Based on Improved HFACS

3.1 Relationship Between Causative Factors of Railway Accidents And Association Rules

The railway system is a comprehensive system composed of many links. The accidents are often caused by the interaction of some links. The occurrence of accidents includes a variety of factors, some of which are closely related, and when problems occur in some factors, the chain reaction will lead to problems in other factors, which will lead to the occurrence of a specific accident. Therefore, the association rules can be applied to the analysis of the causative factors of railway accidents, and the correlation between the causative factors of relevant accidents can be mined out. Through the visualization of association rules, the relationship between the causes of railway accidents can be visually displayed, and the key factors leading to railway accidents can be found out.

3.2 Association Rules

Association rules are a method to mine the relationship between variables in a data set. Related definitions of association rules are as follows:

Definition 1: Item and itemset.
Assume that D is the data set of railway accidents, that is, the transaction set; I is the set of all the cause factors of railway accidents in D, that is, the itemset; T is all the cause

factors of a certain railway accident. One or more items in each transaction are included in itemset I, namely T ∈ I.

The expression form of association rules is A = > B, where A and B are both contained in I, and A ∩ B = Ø, A is the Antecedent, and B is the Consequent.

Definition 2: Support and confidence
Usually, support and confidence are used as the measurement standards of association rules. For itemset A, if Count (A) is equal to all transaction sets containing itemset A; at this time, the support of A is:

$$Support(A) = \frac{Count(A)}{|D|} \tag{1}$$

Similarly, A = > B's support for the number of transaction contains both A and B Count (A, B) divided by the total number of transactions |D|.

$$Support(A \Rightarrow B) = \frac{Count(A, B)}{|D|} \tag{2}$$

At this moment, the support indicates the probability that the item set A and B appear together. For the association rule A = > B, the confidence level Conf(A = > B) refers to the ratio of the itemset containing A and B to the containing item set A in the total transaction set D.

$$Confident(A \Rightarrow B) = \frac{Support(A \Rightarrow B)}{Support(A)} \tag{3}$$

Confidence indicates the probability of including B under the premise of including A.

Definition 3: Lift Since the confidence only considers the support of the antecedents of the rules and does not consider the support of the Consequents of the rules, there will be misleading association rules. Therefore, lift is introduced to remove misleading association rules.

Lift is the ratio of the probability of including B if including A to the probability of occurrence of B in transaction set D.

$$Lift(A \Rightarrow B) = \frac{P(B|A)}{P(\overline{P})} = \frac{Confident(A \Rightarrow B)}{Support(\overline{B})} \tag{4}$$

If Lift > 1, A and B are positively correlated; If Lift < 1, then A and B are negatively associated and these negative association rules are removed. The higher lift, the greater the influence of A and B.

Definition 4: Frequent itemsets
The minimum support, namely Min-sup, is the measurement standard set by the user. If the support of A itemsets is not less than the minimum support, then A itemsets are called frequent itemsets. If B itemsets are contained in A itemsets and A itemsets are frequent itemsets, then B itemsets are frequent itemsets; if B itemsets are included in A itemsets and B itemsets are not frequent itemsets, then A itemsets not frequent itemsets.

3.3 Apriori Algorithm

Apriori algorithm is a common algorithm for association rules. The mining steps of the algorithm are divided into two parts: finding out all frequent itemsets and generating association rules. The specific process of the algorithm: (1) Find out all frequent item sets: it is an iterative method of searching for exclusion layer by layer to find out all frequent item sets. There are two steps: first, the transaction set is scanned to determine the occurrence times of item sets containing the same element, and the itemsets that do not satisfy Min-sup are removed. Second, iterate until no maximum itemsets appears. For example, in the K step, the K-1 item set obtained in the K-1 step generates the candidate k-frequent set. The transaction set is scanned to determine whether the support degree of the candidate itemsets K-1 is greater than Min-sup, and the itemsets less than the minimum support is removed to find the k-frequent itemsets. The above is the connection step and pruning step.

(2) Generate association rules: the frequent itemsets mined in the previous step is used to set the minimum confidence Min-conf to mine association rules. The pseudocode of frequent itemsets found by Apriori algorithm is shown in Table 1.

Table 1. The pseudocode of frequent itemsets found by Apriori algorithm

Input:transaction set D, minimum support Min-sup
L_1 = find_frequent_1_itemsets(D)
For(k = 2;$L_{k.1}$! = ∅;k + +){
C_k = Apriori_gen(L_{k-1})
For each transction t ∈ D{
C_t = subset($C_{k,t}$)
For each candidate c ∈ C_t
c.count + +
}
L_K = {c = C|c.count ≥ Min_sup
}
Return L = $U_k L_k$

CK is the set of candidate item sets of length k, and LK is the set of frequent itemsets of length k.

3.4 Case Analysis

3.4.1 Establish the Database of Railway Accident Causing Factors

Based on the improved HFACS model in this paper, 504 railway accidents from 2008 to 2009 were coded and analyzed. Taking "Guangzhou East Railway Station D725 train for preparing to enter the general C category accident of trains" as an example to classify and code the causes of the accident. It can be determined that this causal factor is coded as "1"; otherwise, it is "0". Table 2 shows the classification and coding of the case report of Guangzhou East Railway Station D725 train for preparing to enter the general C category accident of trains".

Table 2. The classification and coding of the case report of Guangzhou East Railway Station D725 train for preparing to enter the general C category accident of trains"

Improve the classification of HFACS causative factors	Coding
Inadequate safety training X1	1
Insufficient business skills X2	1
Irregular working hours X3	0
Irregular workplace X4	0
Excessive job requirements X5	0
Wrong instruction/execution error X6	0
Communication failure X7	1
Insufficient education of safety responsibility T8	1
Inadequate safety inspection T9	1
Lack of rules and regulations T10	0
Unclear division of job responsibilities T11	0
The lack of safety responsibility consciousnessR12	1
Poor health R13	0
Insecurity R14	0
Insufficient capacity of operatorsR15	1
Equipment design defectsR16	0
Equipment ineffectivenessR17	0
Underutilization of equipmentR18	0
Bad weatherR19	0
Foreign object invasionR20	0
IrregularitiesO21	1
Improper operationO22	1

As can be seen from Table 1, this case can be represented by a Boolean matrix of 1 × 22 dimensions, as shown in Formula (5):

$$\left[1\ 1\ 0\ 0\ 0\ 0\ 1\ 1\ 1\ 1\ 0\ 0\ 1\ 0\ 0\ 0\ 0\ 0\ 1\ 1\right] \quad (5)$$

Similarly, a 1 × 22-dimensional matrix of 504 cases of all railway accidents can be obtained. Finally, all the Boolean matrices of 1 × 22 dimensions are combined into a Boolean matrix of 504 rows × 22 columns, namely the railway accident data set RA-D, as shown in Formula (6):

$$RA - D = \begin{bmatrix} X_{1,1}\text{L} & R_{1,12} & \text{L} & O_{1,22} \\ \text{MO} & \text{M} & \text{O} & \text{M} \\ X_{504,1}\text{L} & R_{504,12}\text{L} & O_{504,22} & \end{bmatrix} \tag{6}$$

This data set clearly represents the causative factors of each railway accident, which provides a basis for mining association rules later.

3.4.2 Mining the Relationship Between the Factors Causing Railway Accidents with Small Data

The Apriori algorithm is applied to the data set RA-D composed of Boolean matrices, where each row represents an accident case and each column represents an accident causative factor item. Through the iterative test, the minimum support is 0.03, the minimum confidence is 0.1, and the minimum lift is 1. For the sake of analysis, only consider the maximum Antecedent term to be 2. Finally, a total of 211 rules were generated from the data set of the causes of railway accidents. Among them, 91.8% of the rules have a support between 0.03 and 0.09. {Improper operation} = > {The lack of safety responsibility consciousness}" has the largest support, with a value of 0.11; There is 82.6% confidence between 0.1 and 0.8, and the greater the confidence, the fewer the rules, "{The lack of safety responsibility consciousness, Inadequate safety training} = > {Inadequate safety inspection}" the highest confidence, the value is 1. At the same time, 79.4% of lift were between 1 and 11. Table 3 shows the top 5 association rules of lift.

Table 3. The top 5 association rules of lift

Aules	Support	Confident	Lift
{Wrong instruction/execution error, Irregularities} = > {Unclear division of job responsibilities}	0.037	0.354	13.528
{Improper operation, Communication failure} = > {Insufficient capacity of operators}	0.055	0.461	13.006
{Inadequate safety inspection, Communication failure} = > {Improper operation}	0.0408	0.381	12.803
{Inadequate safety inspection, Insufficient capacity of operators} = > {Improper operation}	0.037	0.400	12.803
{Bad weather} = > {Inadequate safety inspection}			

The advantage of association rules lies in the visualization of association rules after mining the causative association rules of railway accidents. The close relationship among the causative factors of railway accidents can be shown more intuitively through images, and then further solutions can be proposed. In order to make the image easy to analyze, the

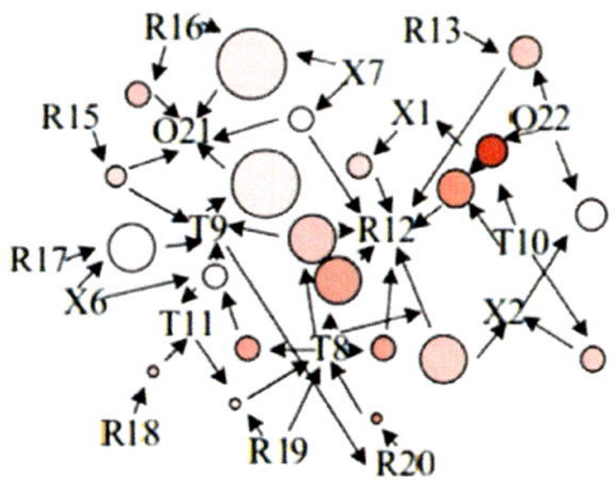

Fig. 2. A visualization of the top 20 association rules sorted by support

first 20 strong association rules are visualized in this paper. Figure 2 shows a visualization of the top 20 strong association rules sorted by support.

It can be seen from Fig. 2 that among these 20 rules, the four railway accidents that cause O21 irregularities, T9 inadequate safety inspection, R12 the lack of safety responsibility consciousness, and T8 insufficient education of safety responsibility are in the middle of the visual image. Five rules point to irregularities, four rules point to inadequate safety inspections, six points to the lack of safety responsibility consciousness, and four points to insufficient education of safety responsibility. It shows that the above four accident-causing factors are closely related to other accident-causing factors and have a high degree of support, so they can be determined as the four source factors, which basically run through all railway accident cases. When these four source factors exist, it is very likely that unsafe behaviors, confusion and inadequacy of management, and related unsafe psychology will occur before and during train operation, leading to accidents. For example, The lack of safety responsibility consciousness often leads to improper operations. The occurrence of Foreign object invasion is often accompanied by factors such as inadequate security inspections. Therefore, taking corresponding measures for these four types of railway accidents can effectively reduce the occurrence of railway accidents from the source and ensure the safety of railway operation. The lack of safety responsibility consciousness, as a relatively important one of all key factors, occupies an extremely important position in the prevention and control of railway accidents. Therefore, relevant railway departments should strengthen safety education, improve corresponding rules and regulations, and give deep criticism to those responsible for safety. Only when safety education is implemented can the safety awareness of all parties be improved and the occurrence of accidents can be reduced.

4 Conclusion

(1) Based on the HFACS classification of the causes of aviation safety accidents, combined with the reality of the railway industry, the original HFACS model was improved, and a new HFACS model conforming to the railway field was established. Raise technical factors to a relatively important position. And under the improved HFACS classification framework, the collected 504 railway accident cases were coded, and the railway accident data set RA-D was established.

(2) Use Apriori algorithm of the association rules to data mine the causes of railway accidents, and visualize part of strong association rules set. Four major factors were discovered, which are irregularities, inadequate safety inspection, lack of safety responsibility consciousness, and insufficient education of safety responsibility. These four accident-cause factors are closely related to other causes of accidents, and are the root factors of other causes. Propose corresponding improvement measures for the lack of a strong sense of safety responsibility for the key cause of the accident.

References

1. Li, C., Tang, T., Chatzimichailidou, M.M., Jun, G.T., Waterson, P.: A hybrid human and organisational analysis method for railway accidents based on STAMP-HFACS and human information processing. Appl. Ergon. **79** (2019)
2. Punzet, L., Pignata, S., Rose, J.: Error types and potential mitigation strategies in Signal Passed at Danger (SPAD) events in an Australian rail organisation. Safety Sci. (2018)
3. Qingjian, Z: HFACS-RAs based railway accident casual factor modeling and hybrid learning approach. Beijing Jiaotong University (2017)
4. Ruth, M., David, G., Richard, M.: Application of human factors analysis and classification system (HFACS) to UK rail safety of the line incidents. Accid. Anal. Prev. **97** (2016)
5. Hardianto, I., Fitri, I.Z.: Indonesian railway accidents--utilizing human factors analysis and classification system in determining potential contributing factors. Work (Reading, Mass.) 41 Suppl 1 (2012)
6. Guochen, Z.: Research on EW-LDA railway accident contributor based method. Beijing Jiaotong University (2019)
7. Liu, Y., Liu, Y., Ma, X., Qiao, W.: A comprehensive model for human factors evaluation in maritime accident: HFACS and FAHP. In: International Informatization and Engineering Associations. Proceedings of 2019 2nd International Conference on Financial Management, Education and Social Science(FMESS 2019), pp. 247–253. International Informatization and Engineering Associations: Computer Science and Electronic Technology International Society, China (2019)
8. Ruiwei, C.: Hazard identification on operation system of high-speed railway. Southwest Jiaotong University (2014)

Baby Carriage Detection in Subway Stations Based on YOLOv3

Xiaoqing Zeng[1](✉), Siyu Ai[1], Jicheng Huang[2], Zhongzhen Ma[3](✉), and Gu Min[4]

[1] The Key Laboratory of Road and Traffic Engineering, Ministry of Education, Tongji University, No. 4800 Cao'an Road, Shanghai 201804, China
1823332464@qq.com

[2] Shanghai Metro Line 17 Development Corporation, Shanghai, China

[3] Shentong Metro Construction Croup Co., Ltd., No. 1283 Yishan Road, Shanghai, China
2959244421@qq.com

[4] Shanghai Municipal Engineering Design Institute (Group) Co., Ltd., Shanghai 200092, China
gumin1@smedi.com

Abstract. In recent years, computer vision technology has developed rapidly and has been applied in various fields. Using this method to detect baby carriage in subway stations can provide technical support for the improvement of future subway station facilities layout. This paper uses the baby carriage detection model to detect the baby carriage that appears in the surveillance video of the subway station. Due to the dense flow of people in subway stations, the frequent flow of people, and the low pixels of the video surveillance system, there are some difficulties in real-time detection of baby carriage. Through the collection and production of the baby carriage data set, the training of the detection model and the final detection verification, we can achieve real-time detection of the baby carriage. The accuracy rate can reach more than 90% .

Keywords: Target Detection Algorithm · Subway Station · YOLOv3

1 Introduction

Target detection has been the research hotspot in the field of computer vision and machine learning for many years. It has also been widely used in many areas, such as transportation, medicine, geography, agronomy, etc., and has made great contributions to solving some difficult problems in such fields. There is no doubt that target detection will play an increasingly critical role in more fields.

Target detection methods can be divided into two kinds, traditional methods and deep learning methods. Traditional methods generally include Viola-Jones, Histogram of Oriented Gradient (HOG), Deformable Parts Model (DPM), Non-Maximum Suppression (NMS) and so on. Xinchen Xu et al. [1] analyzed the features of the human skeleton to detect abnormal behavior event(fall) in traffic video. Xiong Zhang et al. [2] combined the edge features of the color image with the features of HOG Depth image in the pedestrian detection algorithm. The effect of the algorithm in the case of interfering

X. Zeng et al. (Eds.): ITASC 2022, LNEE 1042, pp. 205–212, 2023.
https://doi.org/10.1007/978-981-99-2252-9_19

factors such as illumination, noise, similar colors and obstruction is well. Dongbing Zhang [3] proposed a vehicle target detection method according to color fusion DPM to enhance the vehicle detection rate and reduce the false-positive rate.

Zhiqiang Hou et al. [4] used Generalized Intersection over Union (GIoU) into the NMS algorithm in order to solve the problem of object missing and object false detection in a single threshold NMS algorithm. However, these methods still have the advantages of low detection efficiency and poor detection accuracy,

With the advent of the cloud computing era, target detection has changed from the traditional manual feature extraction method to feature extraction based on convolutional neural networks (CNNS). It can be divided into two parts: one-stage methods and two-stage methods. The representative one-stage methods include You Only Look Once (YOLO) and Single Shot Multibox Detector (SSD). Shitian Zhang et al. [5] modified the YOLO structure to realize the function of recognizing and detecting the contour of an image of the object sampled by a camera. Xiang Li et al. [6] proposed an improved SSD that forces on the small-size target detection. The representative twostage methods are Faster region-based convolutional network (Faster RCNN) [7] Mask R-CNN [8], etc.

In this paper, we use the YOLOv3 structure to detect baby carriage in subway stations. The detection result can provide some technical support for the subway station design and planning. The rapidity of the YOLOv3 network can solve the problem of difficulty in capturing baby carriage targets caused by excessive crowds in subway stations. The high accuracy of the YOLOv3 network can meet the requirements of accurately identifying baby carriage even under the low pixel conditions of the subway detection system.

2 Baby Carriage Detection Algorithm Based on YOLOv3

2.1 Development History of YOLO

Although the R-CNN series has a relatively high accuracy rate, the speed is still too low to achieve real-time detection. Hence, the YOLO network came into being.

Different from R-CNN series' method of selecting region proposal, the YOLO network considers target detection as a regression task to solve. This means that the category probability and position coordinate value of the object are directly returned throughout the YOLO network.

Compared with other target detection networks, YOLOv1 has the following advantages: fast detection speed, high detection accuracy and strong migration ability. However, it still has some shortcomings: the bounding box is inaccurate and the small-scale targets are always missed.

Based on the YOLOv1 network, the YOLOv2 network has some improvements. It introduces the Anchor mechanism. The K-means clustering method is used to cluster and calculate a better Anchor template in the training set, which dramatically improves the recall rate of the algorithm. At the same time, it combines the fine-grained features of the image so that the shallow features can be connected with the deep features, which is helpful for the detection of small-scale targets.

In 2018, Redmon made some improvements in YOLOv3 based on YOLOv2. Instead of using the original darknet-19 network structure, darknet-53 is used in the feature extraction part in YOLOv3. It also uses the feature pyramid network structure to achieve multi-scale detection. The classification method uses logistic regression to replace softmax. The accuracy of target detection and the real-time performance can be ensured.

2.2 YOLOv3 Algorithm

The basic idea of the YOLOv3 algorithm is to divide the input image into several grid cells. If the center coordinate of the target to be detected falls in which grid cell, the grid cell predicts the target, and each grid cell predicts three bounding boxes. By decoding the parameters contained in feature maps output by the network, the center coordinates and size of the bounding box can be obtained.

2.3 Structure of the Baby Carriage Detection Algorithm

As shown in Fig. 1, the whole structure of YOLOv3 includes input, backbone feature extraction network, feature extraction enhanced network and multi-scale output.

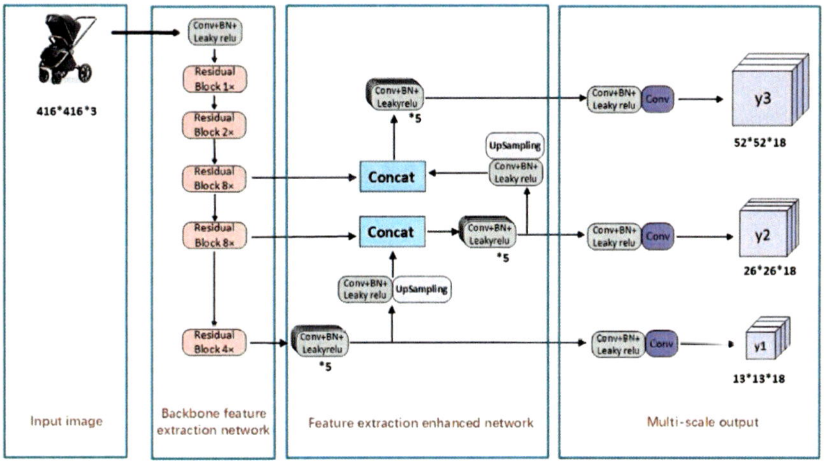

Fig. 1. Structure of the baby carriage detection algorithm

Backbone Feature Extraction Network. The backbone feature extraction network of YOLOv3 uses the Darknet53 network structure instead of the Darknet19 of YOLOv2. The Darknet53 is mainly composed of 1×1 and 3×3 convolutional layers. After each convolutional layer, it contains a batch normalization layer and a Leaky relu layer to prevent overfitting. Conv + BN + Leaky relu is the basic convolution unit of Darknet53. An important characteristic of Darknet53 is the use of Residual, which is easy to be optimized. The internal residual block of Residual uses jump connections to alleviate the problem of gradient disappearance caused by increasing depth in the deep neural network.

Feature Extraction Enhanced Network. This network structure mainly constructs the Feature Pyramid Networks (FPN) to enhance feature extraction. After the input image passes through darknet53, three feature layers will be extracted. YOLOv3 uses these three feature layers to build the FPN. FPN can fuse the feature layers of different shapes, which is beneficial for better features extraction.

Multi-scale Output. Using the FPN, we can obtain three enhanced feature maps, and then the network will convolve these features respectively. The predicted output tensor is $N \times N \times 3 \times (4 + 1 + 1)$. Four represents the adjustment parameter of the anchor box. The first one represents whether an object is contained in the anchor box. The second one represents that the predicted object type is one (in this paper, it refers to a baby carriage). Different from YOLOv2, which outputs five anchor boxes at one time, YOLOv3 obtains three anchor boxes in three dimensions respectively, and obtains the corresponding coordinates and confidence parameters. What's more, YOLOv3 increases the weight parameter. In case of existing an object, weight value will be increased, conversely, it will be reduced. Hence, the loss energy calculation can be more accurate.

Anchor Box. YOLOv3 still uses k-means clustering to determine nine different sizes of anchor boxes as the same as YOLOv2 do. The nine anchor boxes are divided into three groups according to different scales. Each feature map is assigned a group when detecting the target. The yellow boxes in Fig. 2 are anchor boxes. The red boxes are ground truth boxes. The orange boxes are the grids where the center points of the objects are located. It is obvious that 13 * 13 feature map is suitable for detecting large objects. The 26 * 26 feature map is suitable for detecting medium-sized objects. The 52 * 52 feature map is suitable for detecting small objects.

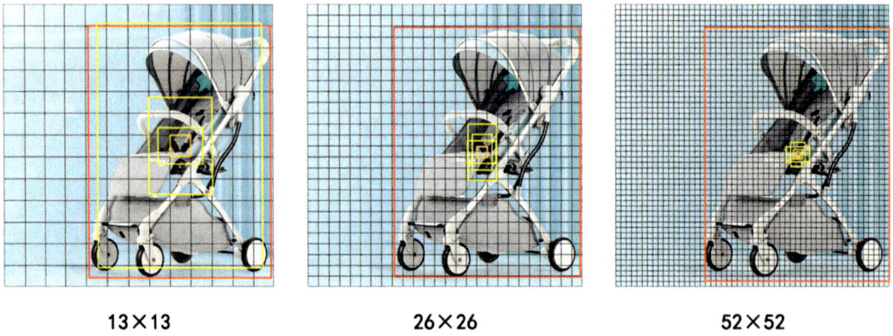

Fig. 2. Comparison diagram of detection box

Bounding Box. YOLOv3 uses a method of predicting relative position directly. It predicts the relative coordinates of the center point of the bounding box, which is relative to the upper left corner of the grid unit. The predicted output is $(t_x, t_y, t_w, t_h, t_0)$. According to the following formula, we can get the coordinates and size of the center of the predicted bounding box.

$$
\begin{aligned}
b_x &= \sigma(t_x) + c_x \\
b_y &= \sigma(t_y) + c_y \\
b_w &= p_w e^{t_w} \\
b_h &= p_h e^{t_h}
\end{aligned}
\tag{1}
$$

3 Experimental Results and Analysis

3.1 Dataset and Preprocessing

The dataset used in the experiment was self-made image dataset of baby carriage, which is mainly collected from the Internet. There are 1273 baby carriage images in the data set, which are divided at a ratio of 9:1, resulting in 1145 images in the training set and 128 images in the validation set. For network training, All of the images are labeled with LabelImg——an image labeling toolkit.

3.2 Experimental Environment and Training

The experimental platform is Windows 10 with Intel(R) i7-8550U and NVIDIA GeForce MX150. The deep learning framework is TensorFlow-GPU 1.13.1. The version of Python is 3.6. The initial learning rate is set to 0.001, and the decay rate is set to 0.92. As is shown in Fig. 3, the loss value has converged after training.

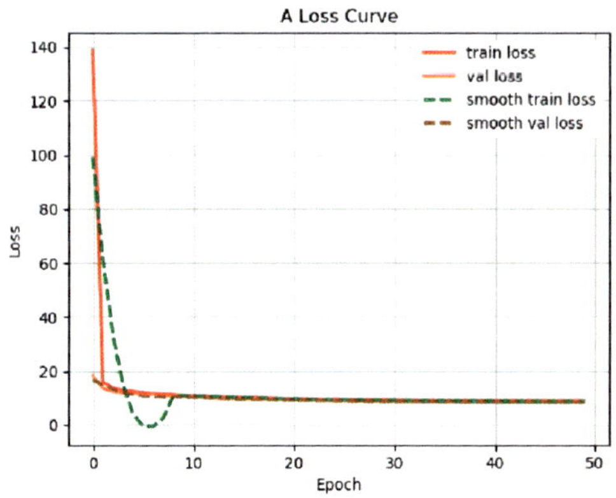

Fig. 3. Loss curve of the model

3.3 Evaluation Indicators

To measure the performance of the model in detecting baby carriage, we selected some evaluation indicators such as Precision, Recall, F_1-Score and Average Precision (AP).

The value of Precision and Recall can be calculated by False Negatives (FN), False Positives (FP) and True Positives (TP), as shown in formula (2) and formula (3). We use 0.5 to divide TP and FP. If the IOU between a bounding box and ground truth is higher than 0.5, it is defined as TP. Besides, it is defined as FP. FN is the situation that the actual object is missed by detection.

$$Precision = \frac{TP}{TP + FP} \tag{2}$$

$$Recall = \frac{TP}{TP + FN} \tag{3}$$

F_1-Score is also known as balanced F_1-Score, defined as the harmonic average of Precision and Recall. The F_1-Score combines the results of Precision and Recall. The value of F_1-Score ranges from zero to one. One represents the best output of the model, and zero represents the worst output of the model. Its specific calculation method is like formula (4).

$$F_1 = 2 * \frac{Precision * Recall}{Precision + Recall} \tag{4}$$

AP is the graphic area enclosed by the PR curve and the X-axis. The definition of AP is shown in formula (5), P is Precision, R is Recall and P(R) is PRC.

$$AP = \int_0^1 P(R)dR \tag{5}$$

The specific values of each evaluation indicators of this experiment are shown in Table 1. Figure 4 shows the curve of each indicator. It can be seen that all indicators are above 95%, which proves that the network has a good effect on baby carriage detection.

Table 1. Table of experimental indicators value

indicators	value
Precision	96.88%
Recall	97.48%
F1-score	0.97
AP	99.70%

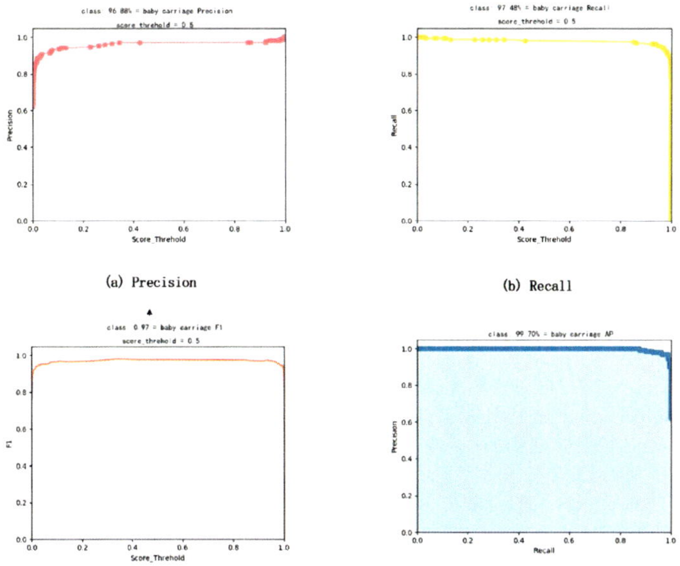

(a) Precision (b) Recall

(c) F_1-score (d) AP

Fig. 4. Indicators curve of experiment

3.4 Detection Result Display

In order to verify the effect of detection model, we intercepted a video which is from Hongqiao railway station of Shanghai Metro Line 17. Figure 5 shows the results of detecting baby carriages in the subway station scene. It can be seen that the model can almost accurately detect the baby carriage that appears in the monitoring, and it can almost guarantee the accuracy and robustness of the detection. The accuracy of the overall detection results can reach an average of more than 90%.

Fig. 5. Results of baby carriage detection

4 Summary

In this paper, we used the baby carriage detection algorithm based on YOLOv3 to detect the baby carriages appearing in the subway video surveillance system, which can achieve the real-time detecting of the target with a high-accuracy rate. The result can powerfully provide some technical support for the subway station design and planning. Due to the limited number of sample sets, the accuracy of the baby carriages detection model needs to be further improved, and further research will be conducted in the future.

References

1. Xu, X., Zeng, X., Wang, Y., Xiong, Q.: Real time gesture (fall) recognition of traffic video based on multi-resolution human skeleton analysis. In: Zeng, X., Xie, X., Sun, J., Ma, L., Chen, Y. (eds.) ITASC 2019. SIST, vol. 127, pp. 1–12. Springer, Singapore (2019). https://doi.org/10.1007/978-981-13-7542-2_1
2. Zhang, X., Shangguan, H., Ning, A., Wang, A., Peng, S.J.A.C., Sciences, C.: Pedestrian detection with EDGE features of color image and HOG on depth images. **54**(2), 168–178 (2020)
3. Zhang, D.: Vehicle target detection methods based on color fusion deformable part model. EURASIP J. Wirel. Commun. Netw. **2018**(1), 1–6 (2018). https://doi.org/10.1186/s13638-018-1111-8
4. Hou, Z., Liu, X., Chen, L.: Object detection algorithm for improving non-maximum suppression using GIoU. IOP Conf. Ser. Mater. Sci. Eng. **790** (2020)
5. Shitian, Z., Guo, Z., Huang, J., Ren, W., Xia, L.: Robotic grasping position of irregular object based Yolo algorithm. In: Editor (ed.) Book Robotic Grasping Position of Irregular Object Based Yolo Algorithm (2020, edn.)
6. Li, X., Luo, H.: An Improved SSD for small target detection. In: Proceedings of 2021 6th International Conference on Multimedia and Image Processing (2021)
7. Ji, S., Han, R., Wei, J., Wang, R.: Clothing image detection and recognition based on faster R-CNN. IOP Conf. Ser. Mater. Sci. Eng. **790** (2020)
8. Fang, Y., Dai, Y., He, G., Qi, D.: A mask RCNN based automatic reading method for pointer meter. In: Editor (ed.) Book A Mask RCNN based Automatic Reading Method for Pointer Meter (2019)

Force Analysis and Details Design of Complex Nodes in Water Supply and Drainage Structures

Gang Wang(✉)

Shanghai Municipal Engineering Design & Research Institute(Group) Co., Ltd., Shanghai 200092, China
wanggang@smedi.com

Abstract. There are various special structures in water supply and drainage engineering. High stress usually appears in engineering design, and directly using the high stress on the structure design may cause diseconomy. Careful analysis is crucial in these special structures. All kinds of structure styles can be adopted, usually using or adding strong nodes is useful, such as the beams, columns or part frameworks in the structures with large internal force, so as to share the large force, leading to redistributing the force. This not only avoids stress concentration, but also avoids adding thickness or volume to the whole structure which may cause unnecessary cost. By several engineering examples, finite element calculation is used to analyze different treating methods. According to the comparison of these calculation results, some special elements such as beams, columns and small thick slabs, are found useful to redistribute the force and optimize the structural design. The methods in this paper may provide reference for water supply and drainage engineering, which can also inspire other engineering projects on complex nodes.

Keywords: water supply and sewerage engineering · internal force redistribution

1 Background

Water supply and drainage structures are important in water supply plants, wastewater treatment plants and pump stations. There are various kinds of structure types, which should firstly meet the water treatment process requirements, so the scale and depth varies according to all kinds of conditions, which are very different from other normal constructions. Meanwhile, the crack-controlling criterion is much stricter because of containing water, so the force conditions should be carefully calculated in some complex working conditions, and special structural measures should be adopted when necessary. By some conventional examples, the special nodes are focused in this paper such as beams, columns and walls, and then special treatments are given by the calculated force results, which may provide reference for similar engineering projects.

2 Panel Constrained by Closed Vertical Framework

When the horizontal span excels the vertical span, and the area of a tank wall is relatively large, this tank wall may be quite thick according to the calculated mid-span moment.

X. Zeng et al. (Eds.): ITASC 2022, LNEE 1042, pp. 213–220, 2023.
https://doi.org/10.1007/978-981-99-2252-9_20

The effect of adding thickness is probably not very good, so a closed vertical framework is usually adopted, consequently, the internal force could be partly transferred to this vertical framework, and then the internal force of the tank wall could be significantly reduced, and as a result, the economy and rationality would be improved. In water supply and drainage engineering, such problems usually appear in the gate well, such as shown in Fig. 1. One of its walls may be tremendously large because it should connect one or several big culverts or pipes. But if a closed vertical framework is designed together with the tank wall, the force may be partly transferred to the framework so as to reducing the wall force, and the thickness and reinforcement as well, which is shown in Fig. 2.

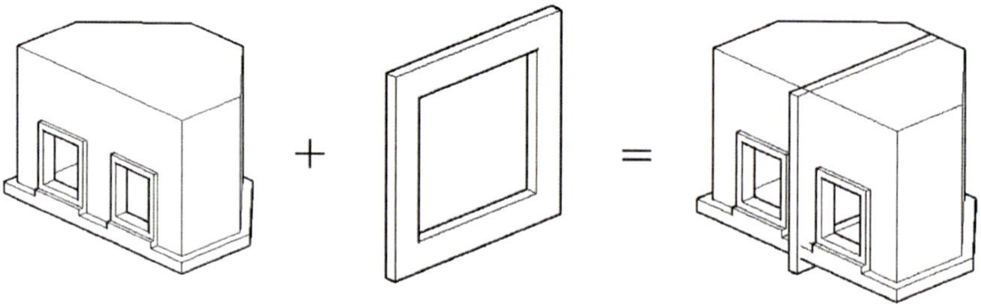

Fig. 1. Closed vertical framework in an inlet gate well.

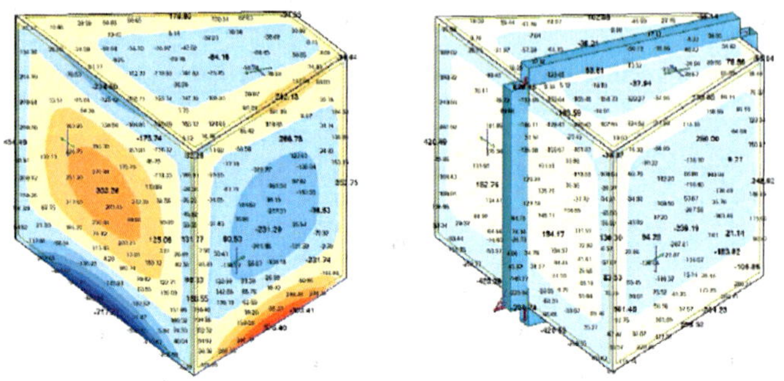

Fig. 2. The impact of closed framwork for structure.

The force of one wall, the top plate and the base plate is quite big in this gate well. According to the calculation results by Autodesk Robot, although the top plate force could be treated by beams, but the internal force of the root of the wall is still big under the outside earth pressure, which is 717 kN*m. It is not economic to use this result for design directly. However, the root moment of this wall is significantly reduced by adding a vertical framework, which is 420 kN*m. And consequently, this wall can be designed as thinner plate with less reinforcement, which is much more economic.

3 Adjacent Structure as Restraints

There are plenty of structure types in water supply and drainage structures. When the internal force of a big plate is also very big, sometimes the treatments could be implemented by using adjacent structure according to the calculation results, and the structure members under low load could be designed much stronger, which would become the restraint of other structure members. It can not only reduce the internal force of long span members, but also enable the existing secondary members to play a favorable role. All the members should be modeled and calculated together during the design, which means the secondary members should be treated as load-bearing elements.

Three examples are taken in this paper for the analysis.

1) Part of an underground wastewater treatment plant is shown in Fig. 3. A channel is hanged outside the tank wall, which is usually only considered under its own pressure of earth and internal water, normally the thickness of 300 mm is enough to meet all working conditions. But in the underwater wastewater treatment plants, the whole structure is relatively deep, which may achieve 15 ~ 30 m, so the earth pressure could cause tremendous internal force in the main tank walls. If the internal force results are directly used for design, the thickness would also exceeds normal structure above the ground, which is very uneconomic. Meanwhile, the horizontal dimension of underground wastewater treatment plant is also very big, it is quite difficult to design closed vertical frameworks, so the holistic constraints cannot be easily formed, and as a result, setting local constraints become an option. Coincidentally, the hanging channel is quite available. If the top and bottom plate of this channel are designed thicker, then they could become constraints to the vertical tank wall as horizontally rotated beams. In fact, if the top, side and bottom plate can be together considered as a rotated box girder, so it can provide bigger constraint effect, and then the internal force of the main wall could be reduced, so as to optimize the thickness and reinforcement (Fig. 4).

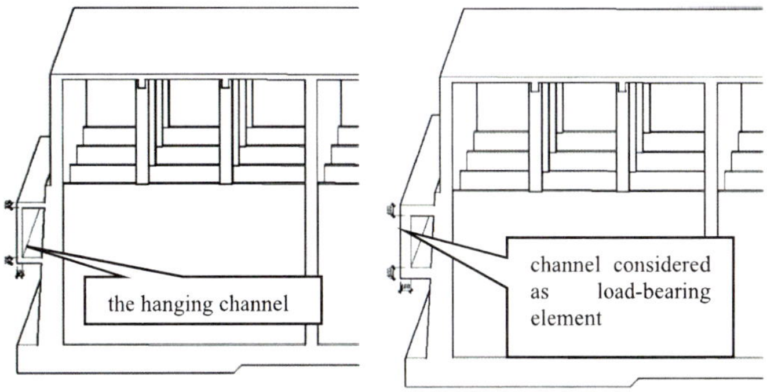

Fig. 3. The outside hanging channel of an underground wastewater treatment plant (thin and thick channel).

By comparison of two models, when the top and bottom plate are modeled as beams, the maximum calculated moment of the main wall is reduced from 1035 kN*m to 645 kN*m, which is almost 40%.

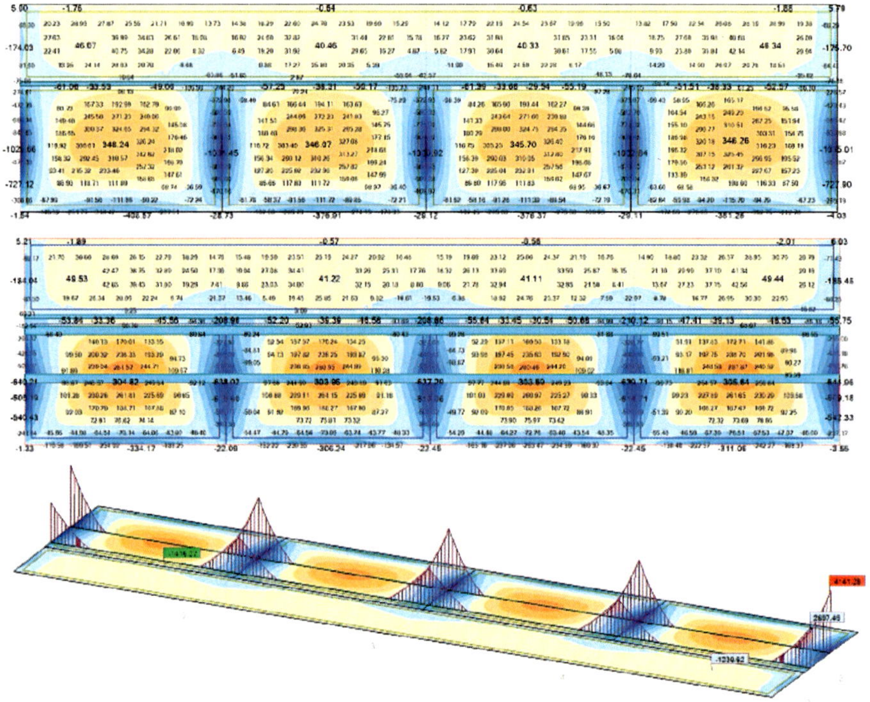

Fig. 4. The comparison of thin and thick hanging channel for restraining wall force.

2) The 2nd example is a rainwater storage pool. There is usual a lower pit in the storage pool, which is used for laying submersible pumps. The lower pit cause obvious impact for the structure force, especially on the corner of these plates. In this condition, treatments could be designed according to the wall shapes, such as setting hidden beams, hidden columns, or local thicker plates. All these treatments are inspired by setting stronger nodes in the parts with relative larger force, making these special structure elements share enough force, so as to avoid adding thickness to the whole tank wall and causing design waste as well (Figs. 5 and 6).

3) The 3rd example is a bottom plate of a circular caisson pump tank. Non-stressed diversion walls are designed in the preliminary design. But the area of the bottom tank is big, when loaded by buoyancy or pure opposite ground force, the mid-span moment is huge. Several convex bottom beams are set in the bottom plate, and then the plate is segmented into small parts, which reduces the internal force. However, the excavation of the earth between the caisson cutting edge and the bottom plate is very difficult, and formwork of these bottom beams are also difficult, so finally another scheme is adopted. In this scheme, the bottom beams are canceled, and the diversion walls are designed stronger, meanwhile, a long deep beam is set to hold these diversion walls, which provides a effective constraint system for the bottom plate. In the model calculation, the maximum moment is controlled under 550 kN*m, which is close to the

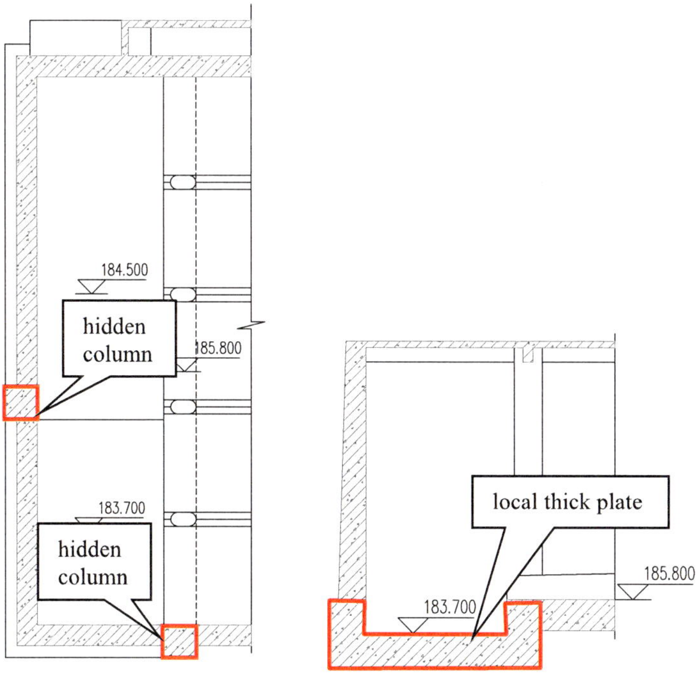

Fig. 5. The structure solution in the pit of a storage pool.

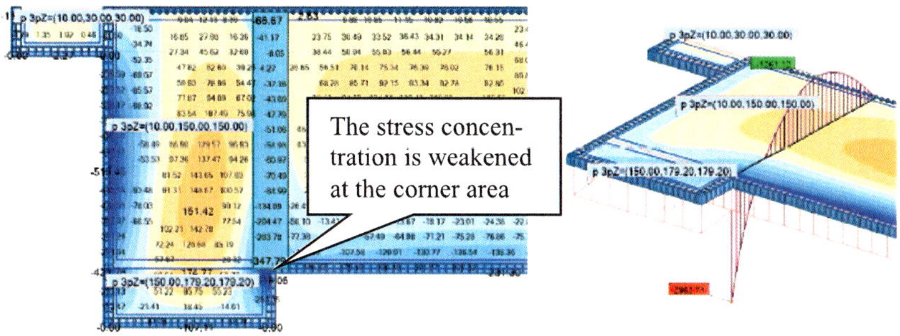

Fig. 6. The force of the storage pool by adding restraining members.

preliminary design with bottom beams, but the second scheme reduces construction difficulty significantly (Figs. 7 and 8).

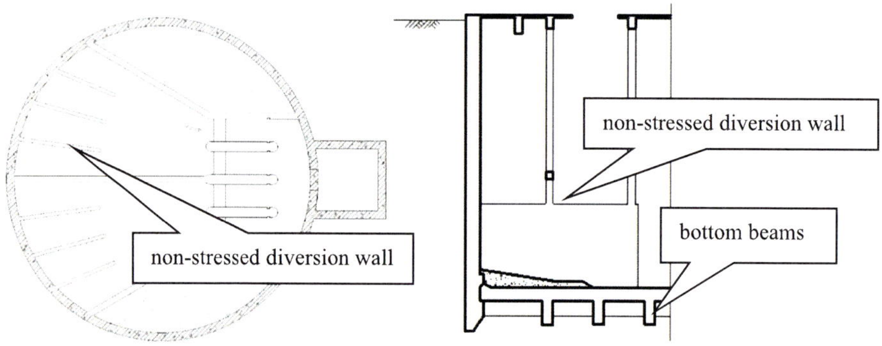

Fig. 7. The layout and section of a circular caisson pump tank.

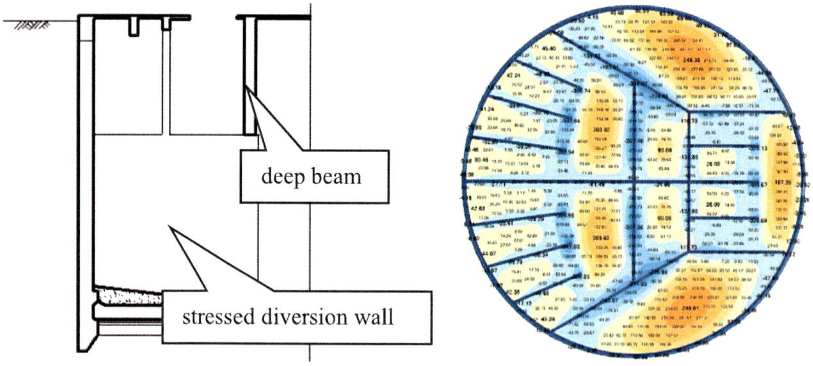

Fig. 8. The raft force when using internal structure as restraints.

4 Treatments Besides the Structural Joints

Long dimension structures are usually designed in water supply and drainage engineering projects. Post-cast strips, reinforcing strips, induced joints and complete joints are usually designed to meet the requirements of tank dimension. Therein, the reinforcement is completely separated by complete joints, so the force cannot be transferred through the joints, but the joints still change force distribution totally, which may cause force concentration in the corner besides the joints. The concentrated force may exceed a lot to other places. It will cause very big thickness if the wall is designed according to the maximum moment, which is not economic or rational. There are many treatment measures on this situation, such as varying walls with bigger thickness in the large force place and smaller thickness in other places, or setting hidden columns or buttress columns, considering the wall as beam and slab structure, so the internal force could by effectively reduced.

An intermediate pumping house of a large wastewater treatment plant is taken as an example. As this pump house is connected with two large pools, and the total dimension is about 150 m, so complete joints must be set. As the pumping house is laid in the middle of two pools, considering the two pools should be symmetric on the outside of the joints, so the joints are set in the pumping house. And more, there should be two joints in this pumping house in order to avoid the gate well. The final scheme of setting joints is shown in Fig. 9.

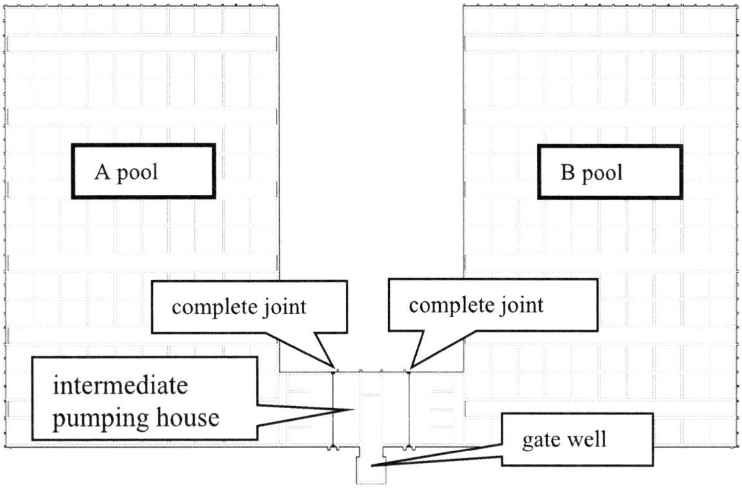

Fig. 9. The layout plan of four tanks of a wastewater treatment plant.

The force of the pumping house is affected by complete joints. Both sides of the joints should be calculated respectively, and meanwhile, there should not be any constraints on the joint. The left part of the joint of this pumping house is taken as an example. According to the calculation, the force of the corner besides the joint is several times of other places, which is 606 kN*m, and the maximum force of other places is only 225 kN*m. If the whole wall is designed according to 606 kN*m, it may be rational to use the thickness of 800 mm, but this would cause most part of the wall exceeds the necessary thickness a lot. During the design process, if a buttress column is set nearby the joint, then the force will all be controlled under 300 kN*m, which only needs thickness of 500 mm, and the reinforcement can also be optimized (Fig. 10).

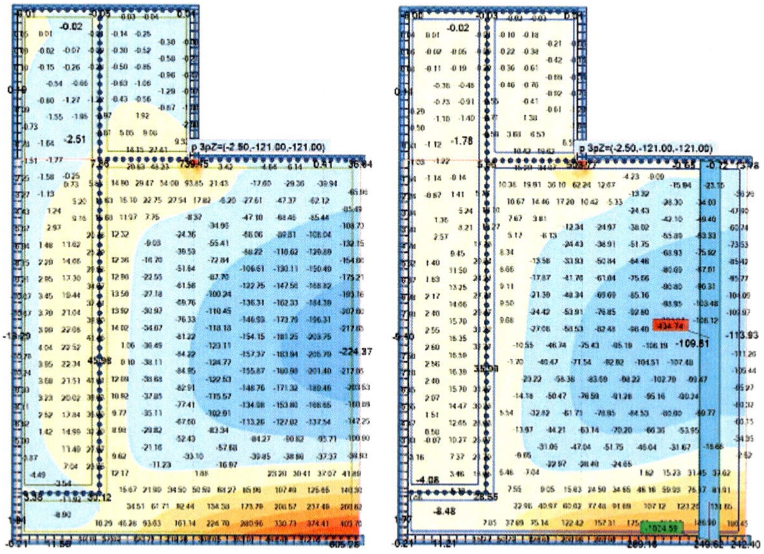

Fig. 10. The comparison of two methods for treating structrue joints (with or without buttress column)

5 Conclusion

There are various kinds of water treatment structures in water supply and drainage engineering, which cannot be concluded by one method. As the scale and depth of these structures increases, the calculated force usually exceeds normal results a lot. Careful analysis is needed by using all kinds of structural measures, such as using the beam-slab structure so as to adding constraints, and then the internal force could be redistributed, which avoids the force concentration. In this way, part of the force is transferred to the local constraining elements, and stronger bending capacity of beams and columns is better used, and consequently, the whole structure is optimized.

References

1. Zhang, D., Xiaoyu, B., Chen, Z.: Discussion on design concepts and methods of thick slab structure in water supply and drainage structures. Special Struct. **37**(2) (2020)
2. Liu, Y., Mao X.: Engineering statistics of thick plate, one-way and two-way slabs. Construction & design for project (2016)
3. Ge, Y.: Three-dimensional static and dynamic comparison analysis of beam slab structure and thick-slab ribbed structure in a power station installation room. Shaanxi Water Resources (2020)

Author Index

X. Zeng et al. (Eds.): ITASC 2022, LNEE 1042, p. 221, 2023.
https://doi.org/10.1007/978-981-99-2252-9